_ FOR MY SONS

[VANISHING]

de.MO

_ TABLE OF CONTENTS

[VANISHING]

essays by ***michael persson***

This book is an insight into what most have no way of seeing.
IN A SENSE IT IS A PAIR OF EYES.

When we ask one another,
HOW ARE THINGS ?
our reply is
a feeling founded on
what we know.

When we answer,
NOT TOO BAD,
that's what
we mean since
that's what we believe
to be true.

What you are about to see
may prompt an altogether different response.
It may shake something inside you ?

THIS PROJECT HAS BEEN A SIXTEEN-YEAR JOURNEY TO SOME OF THOSE UNSEEN PLACES .

The pictures inside show what might never return.
What the eyes have looked upon may be lost forever.
This book is a tour through
several endangered life-forms.
At present, most of their outcomes hang in the balance,
though for some they hang by a thread.

[VANISHING] speaks on behalf of life despite man's ever-threatening presence.
This body of work offers nothing in the way of answers,
neither is it a sermon in hopes of brighter days.
No, its intent is to pose that every-day, run of the mill question.
[VANISHING] gives those who go about their business,
living their lives,
a chance to. look beyond
their worlds and
into others.
Perhaps having closed
the covers on this
uncommon journey
people in future
may respond
differently to that
question
that they so often hear
and that this book asks once more.

for the fate of what might be ...

[VANISHING]

The metal that glistens
on your neck,
those nuggets on your ears,
adorning your nose,
and resting on your fingers are there because of [CYANIDE].

[CYANIDE] leaching is the method of choice to get gold from ore,
to lift away the *prima materia*
from the rock
in which it lies.
[CYANIDE] even lifts gold too small to be seen by the naked eye.

THIS **SUPERTOXIN** REMOVES EVERYTHING OF WORTH AND LEAVES NOTHING , BUT ROCK .

Healthy land poisoned by **[CYANIDE]** extends
outwardly from lunar craters dug to depths men normally don't go.
Hundreds of workers scurry like ants in
these cavities blasting **[CYANIDE]** solution against rock,
carrying ore heavy on their backs,
wading in a toxic soup.

The gluttony for gold by mining companies,
governments and financial institutions alike has reduced the value of what
once set the standard for currencies around the world by a quarter.
[CYANIDE] and **[GREED]**
has increased the volume of this precious metal and at the same time cheapened it.

GOLD WILL SOON BECOME A SIMPLE COMMODITY
SOLD ON TELEVISION BY AN OVER-EAGER BLONDE WITH NICE HANDS ,
A METAL AS WORTHLESS
AS THE GROUND THAT CYANIDE HAS HELPED
LEACH AWAY TO NOTHING .

guyana _ the poisoning of something precious

_ *Section*

I

THE POISONING OF
SOMETHING PRECIOUS

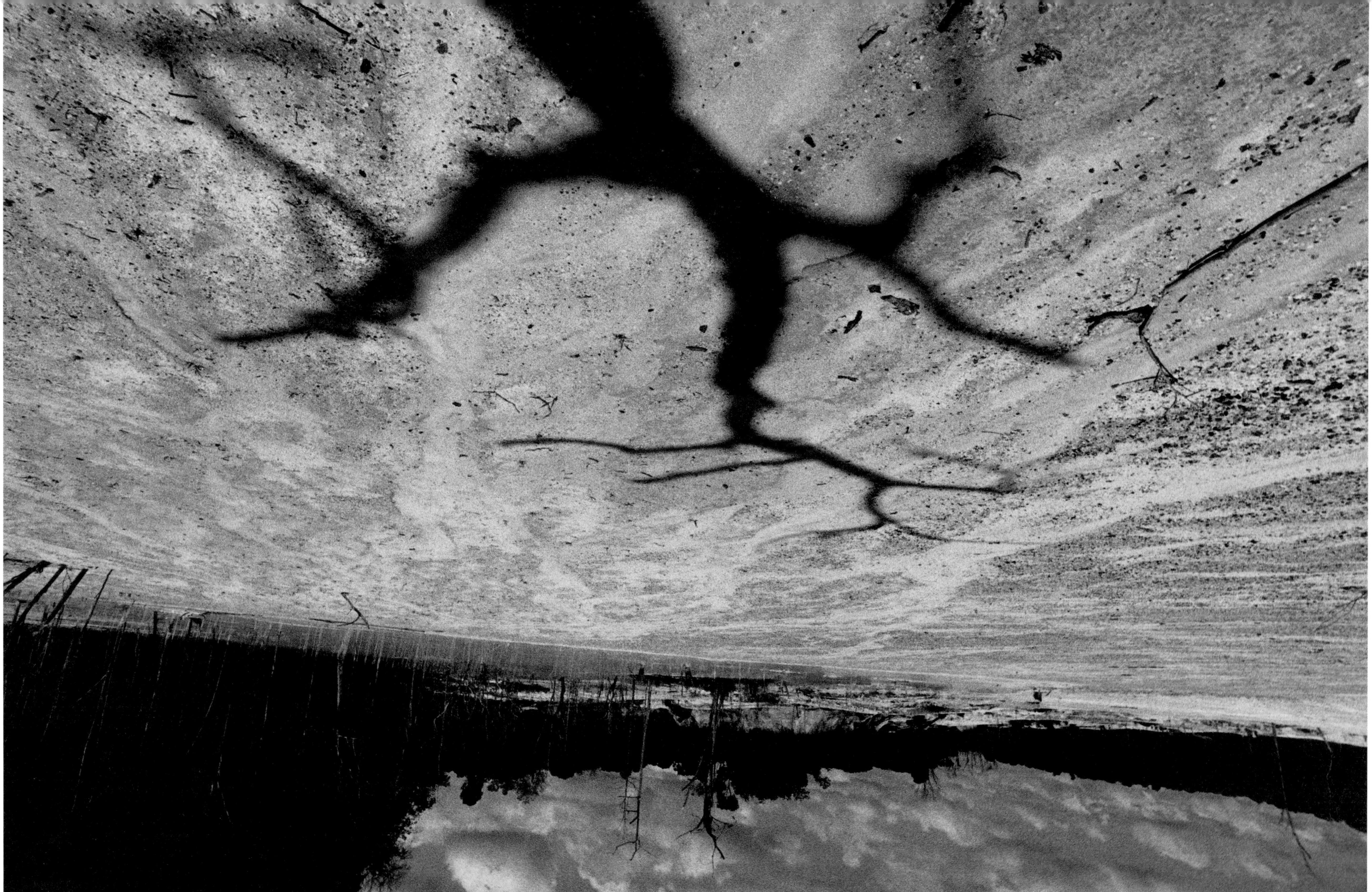

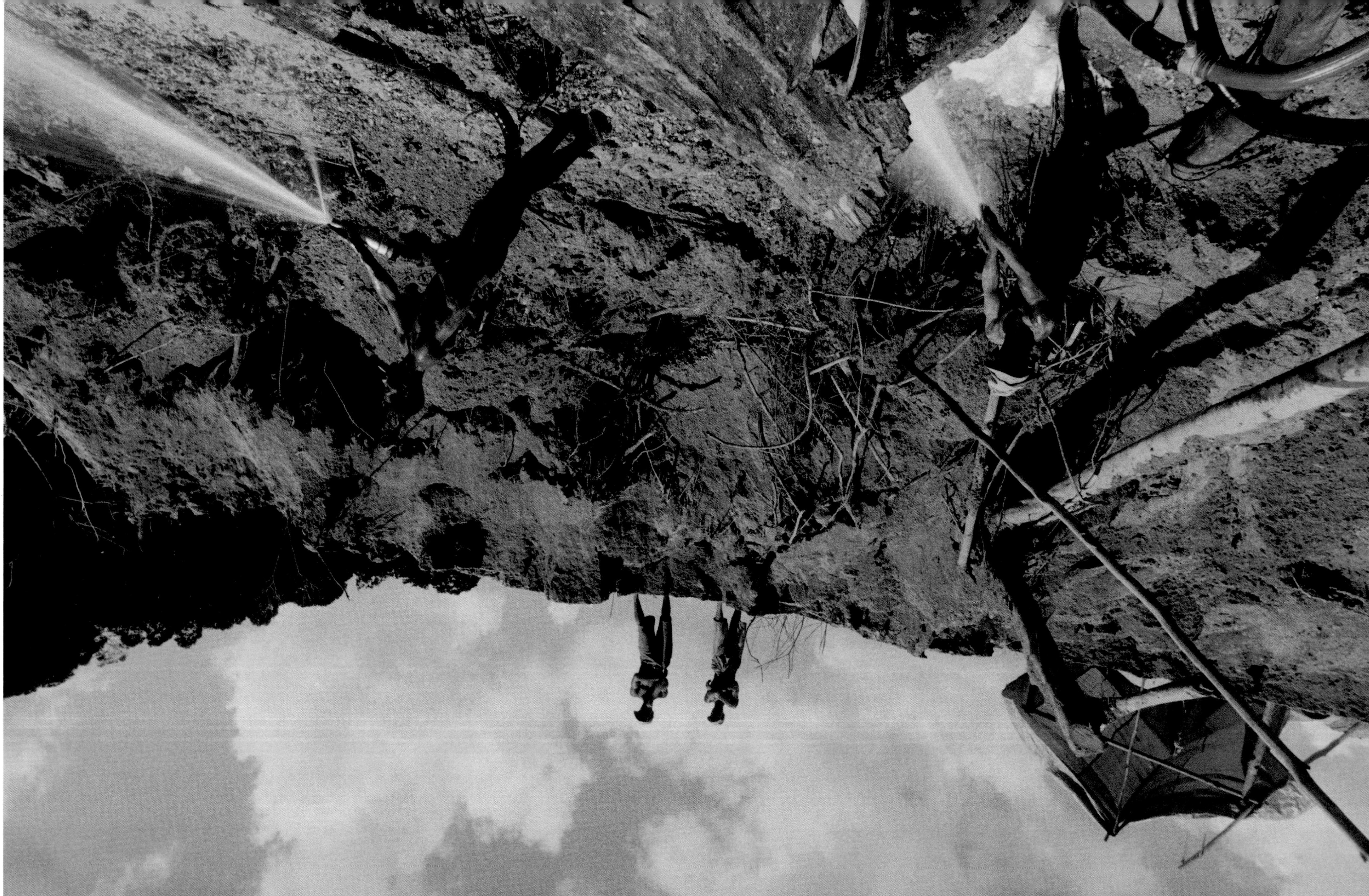

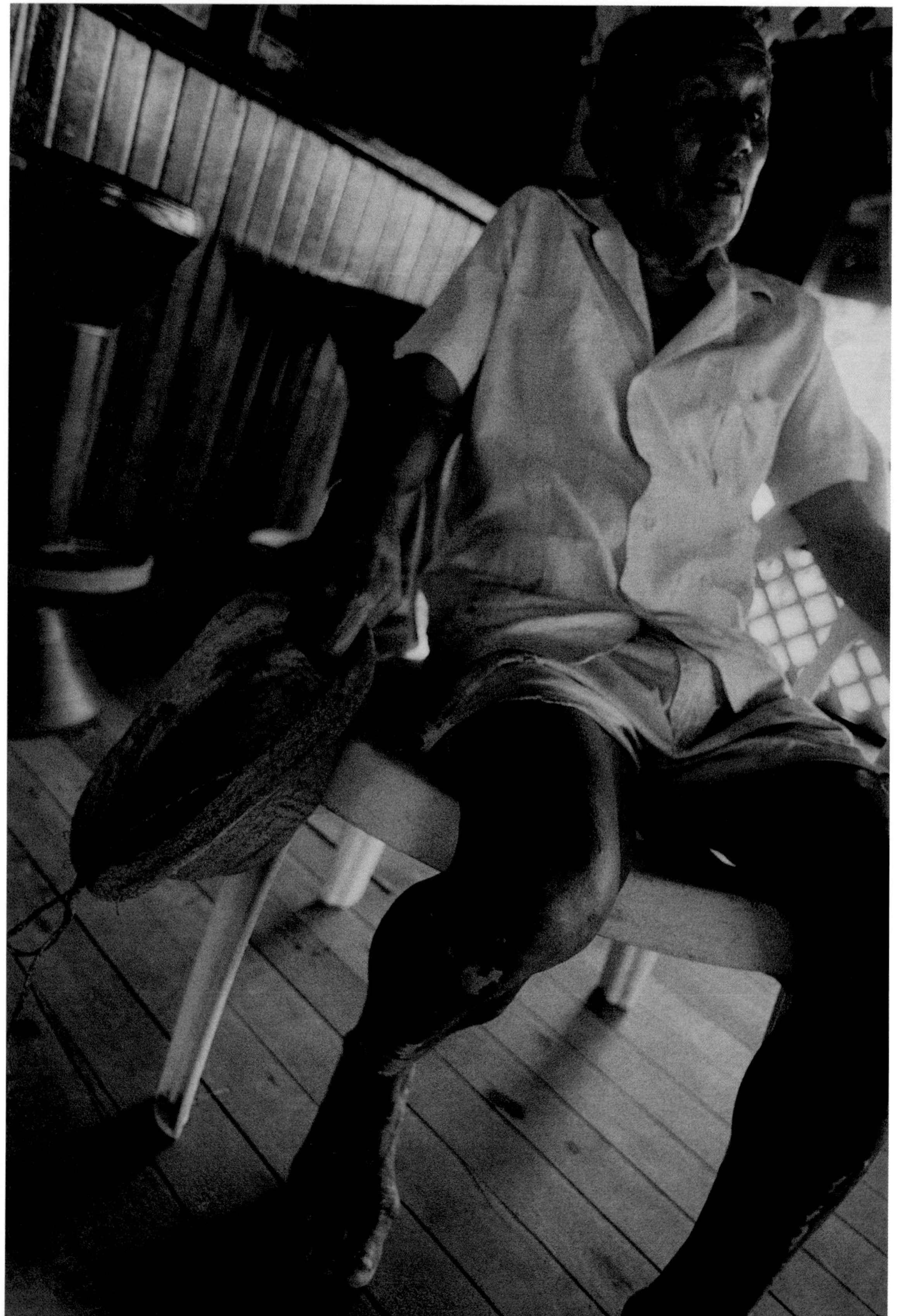

_ 016 SUFFERING CYANIDE'S EFFECTS . MAHDIA
_ 017 COMMERCIAL GOLD MINE WORKERS . OMAI

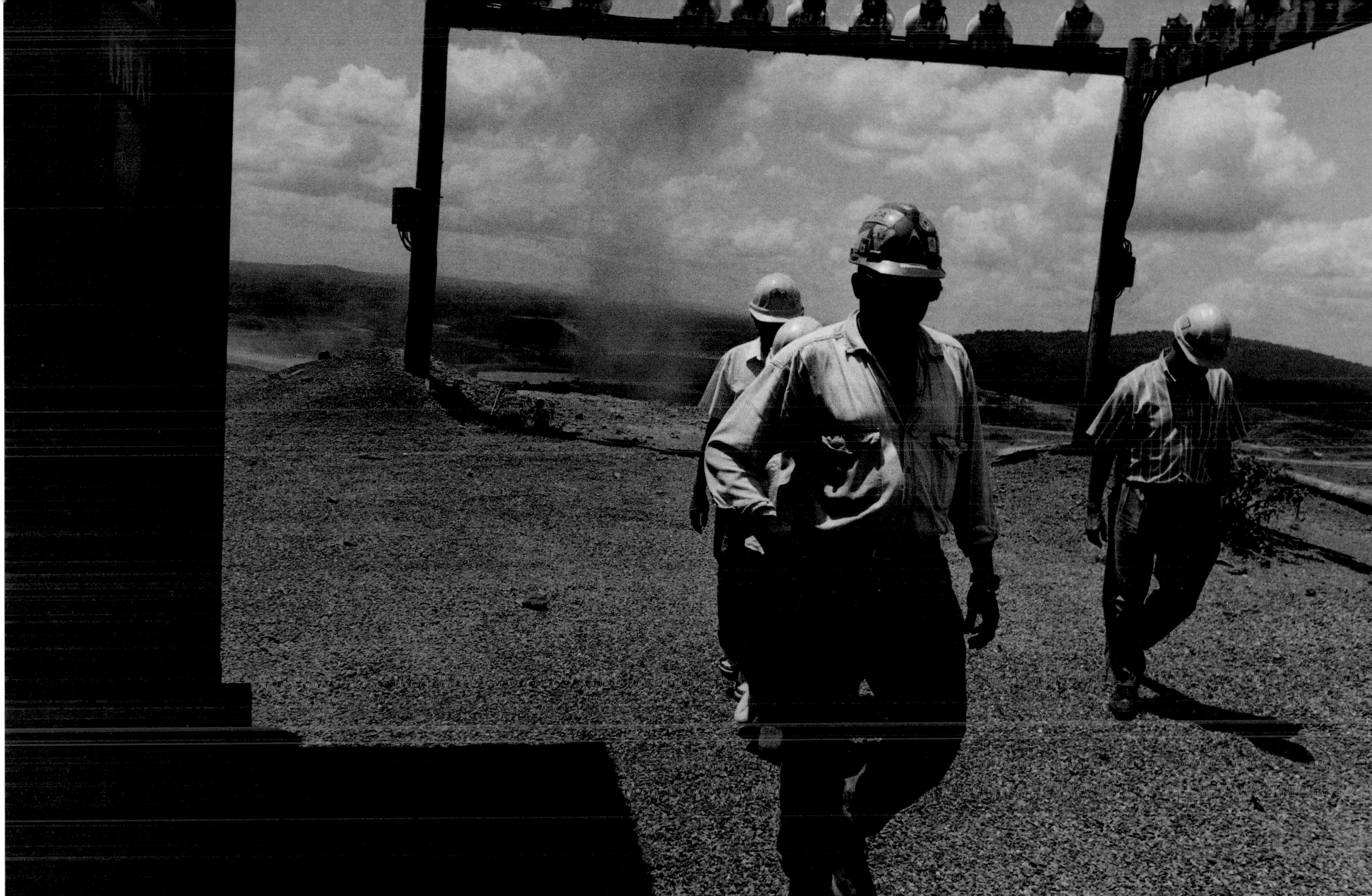

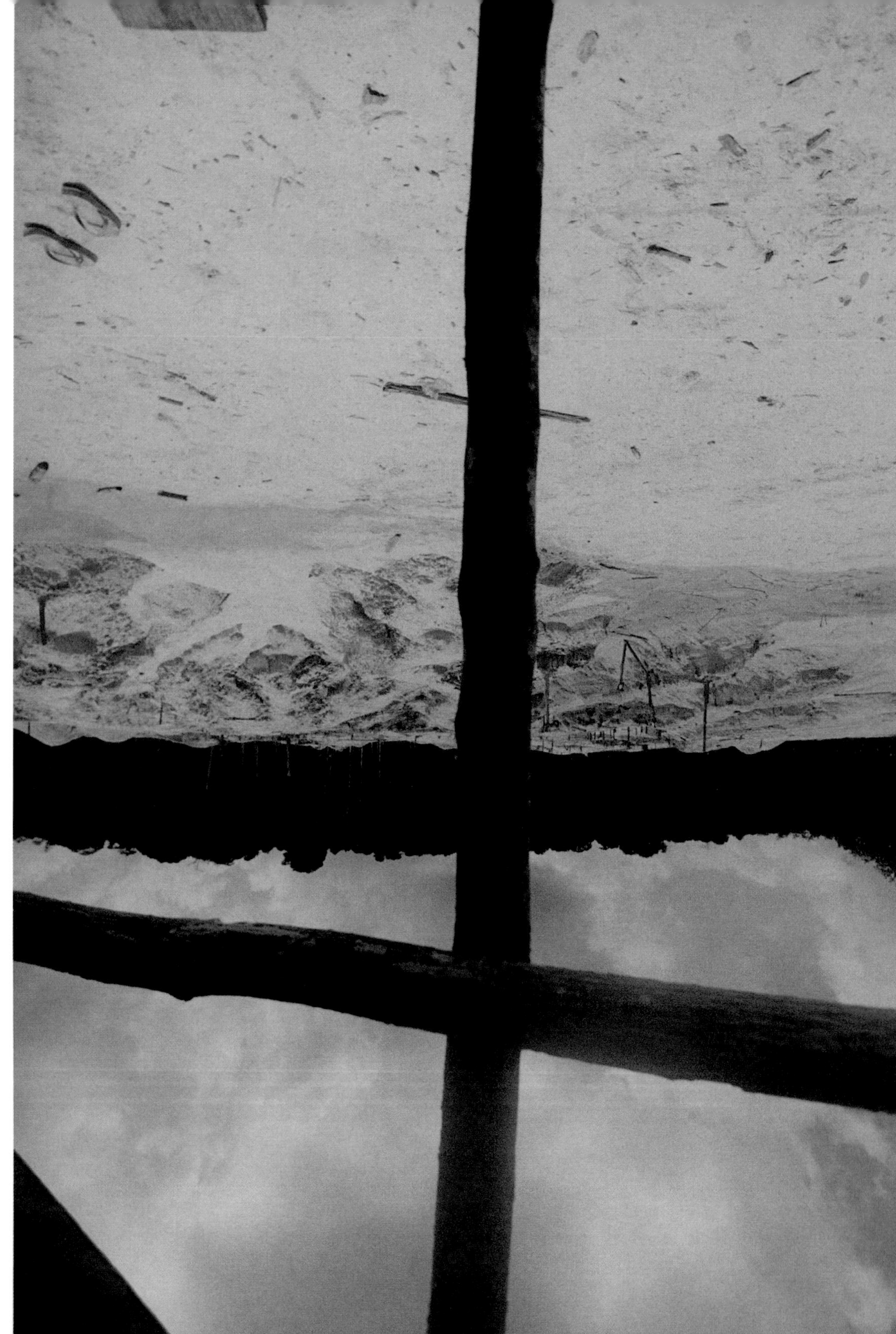

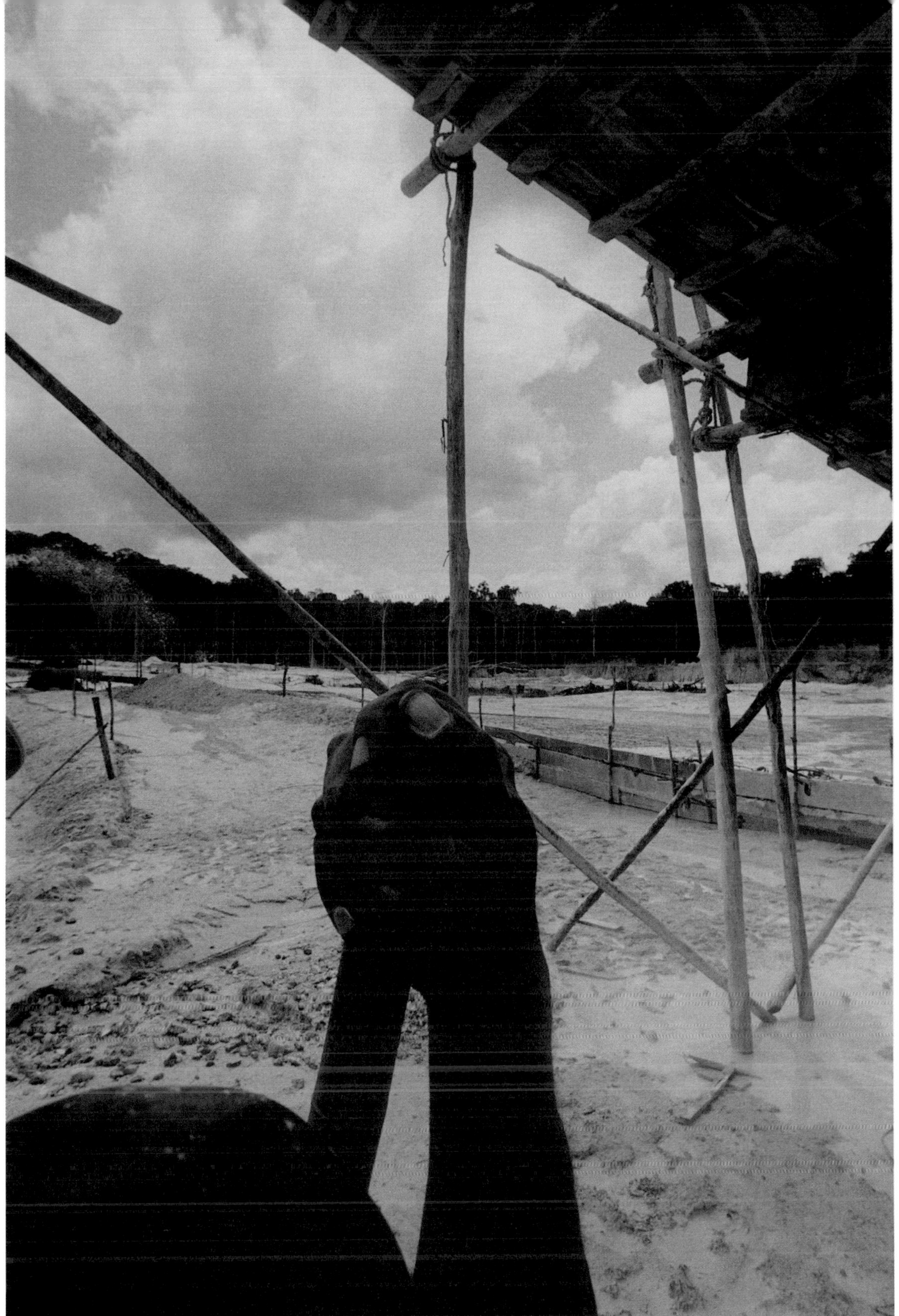

_ 018 POISONED EARTH . MAHDIA
_ 019 GOLD MINIG CAMP . MAHDIA

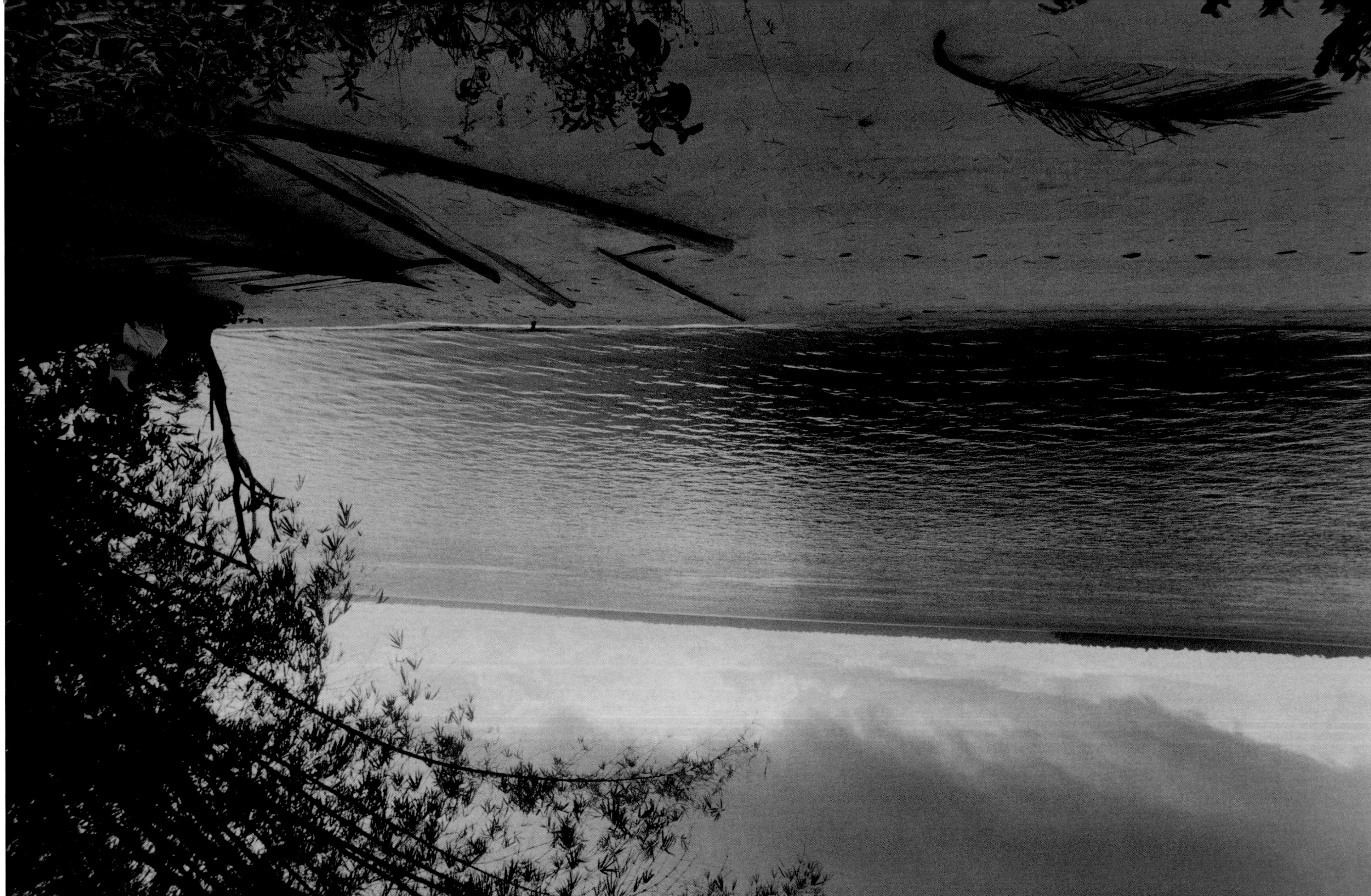

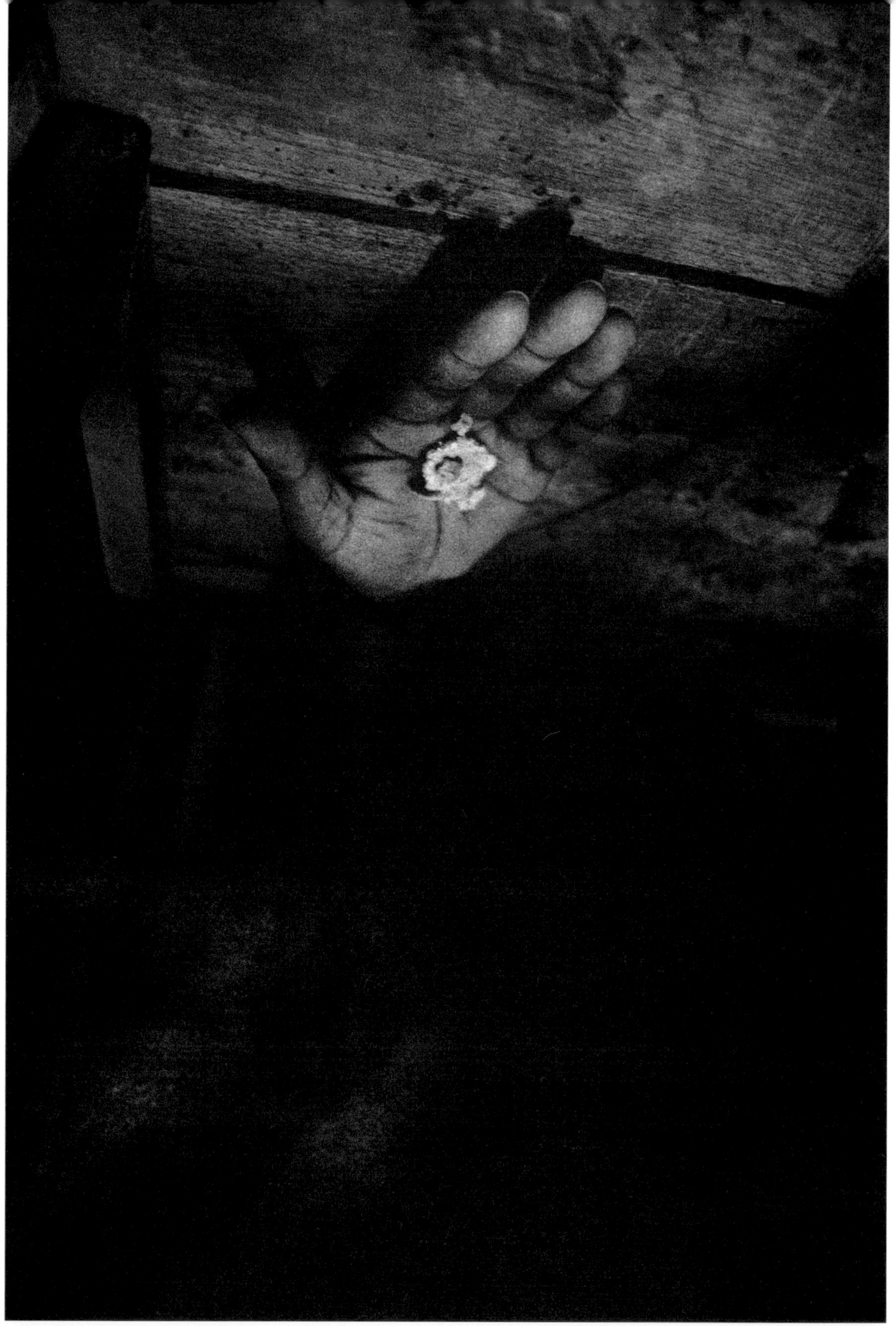

_ 023 GOLD BUYER SHOWS A NUGGET. MAHDIA MINING CAMP

A Mercedes leaves the town of Most in the Czech Republic.
An old woman sits looking out the rear window.
She's looking out and crying.
A fine rain falls and the afternoon, like the town, is grey.
It's grey in the sky,
grey on the ground,
grey in the faces of the people here.
There is no other color should the sun even shine.
The woman looks about 80.
Born perhaps in the early 1920's.
Then, Most had a different name, a different group of inhabitants.
Then, Most was Brüx and Northern Bohemia was the Sudetenland.
The Germans were once the majority here.
They left this place long ago.

HAD SHE LIVED HERE IN THOSE TIMES, SHE MAY HAVE REMEMBERED THIS FILTHY PLACE AS SOMETHING ELSE.

There were rivers that slipped through wooded groves
and pastures where dairy herds dotted the farmlands.
The land was the yoke,
and people were bound to it and their history,
contentedly.
Green fields, clean water, sweet rain.
Now there is acid rain, oily wet, and the air stinks of sulfur.
What trees are left stand spindly, stripped bare.
Not much grows other than the children who grow old before their time.

NOT MUCH SURVIVES.

A DOCTRINE OF SOCIAL ENGINEERING WAS THE CAUSE.
FORTY-FIVE YEARS OF FORCE-FEEDING PEOPLE THEIR FATE ALTERED WHAT WAS RICH INTO A RUIN WHOLLY DERELICT.

The woman looks long and hard through the greasy glass ... her mouth slightly open.
She's slightly agog.
Rapid militarization, heavy industrialization and negligence has emptied Brux of its native beauty ... demolished the yoke.
The German majority had no choice but to leave and then an iron curtain was drawn across.
What had been abandoned was re-settled and the land cut open and its wealth sucked out.

Leaving here, why does the visitor cry?
She's going home.
North, across the border in a Mercedes bearing German license plates.
She has witnessed first-hand what was done behind the curtain.
Could this once have been her home?
Seine Heimat?
Did she come to see the memory?
The place of exiled remembrances?
Is she crying for what she has seen?
Or for her history and her family's history and her peoples' history that has been piled in heaps.
Wiped away.

bohemia _ memories of things lost

_ *Section*

II

MEMORIES OF THINGS LOST

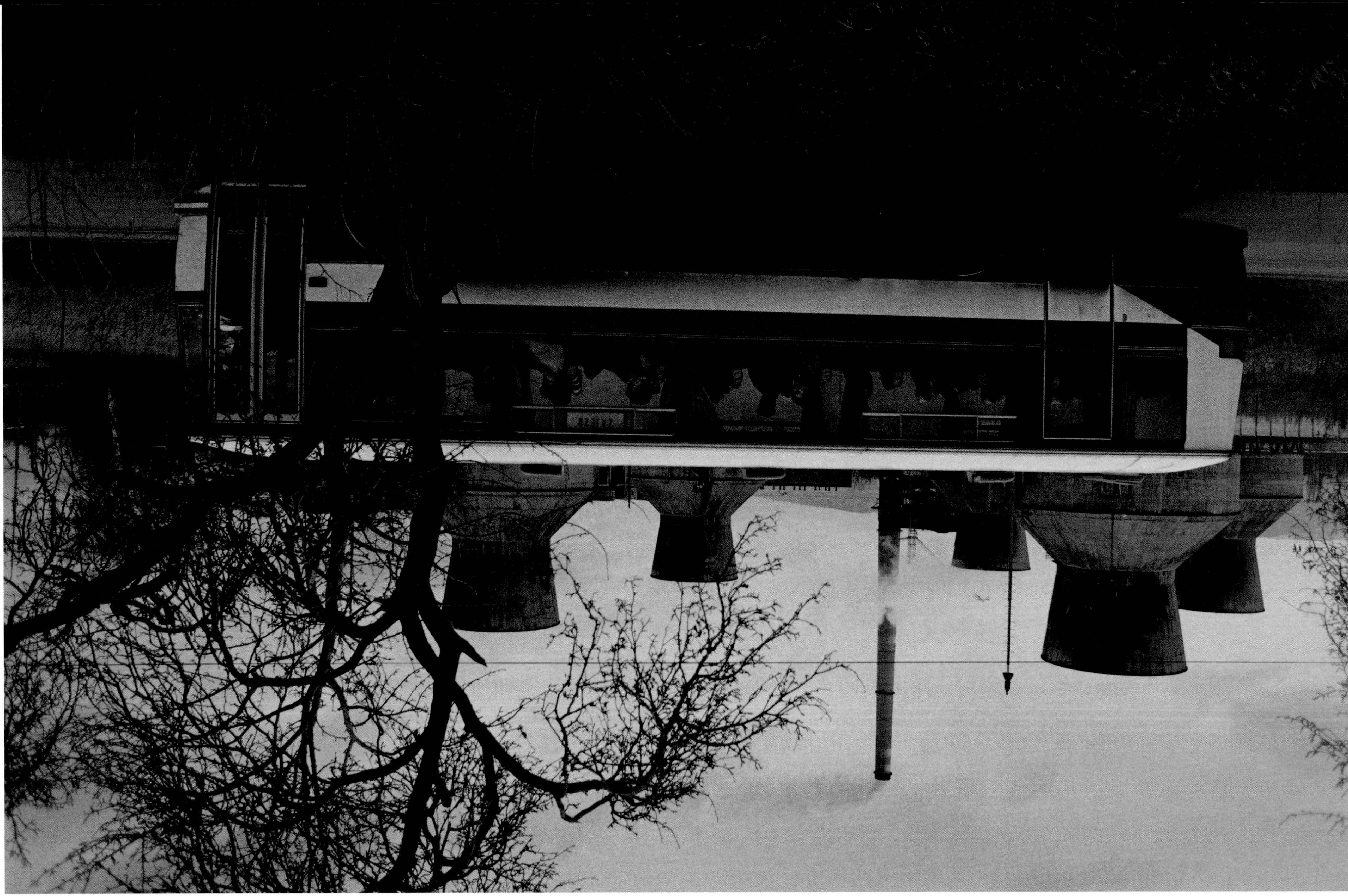

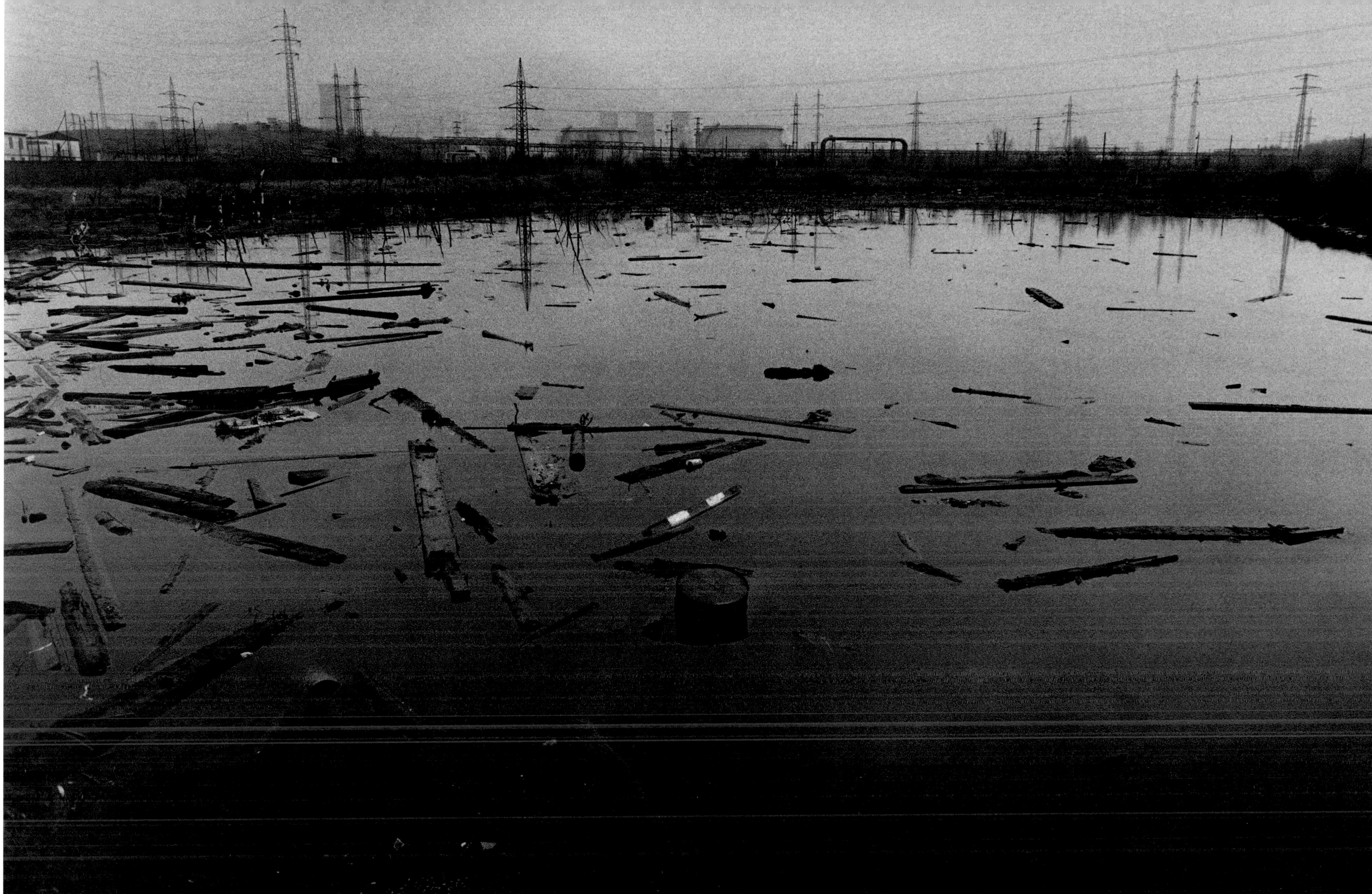

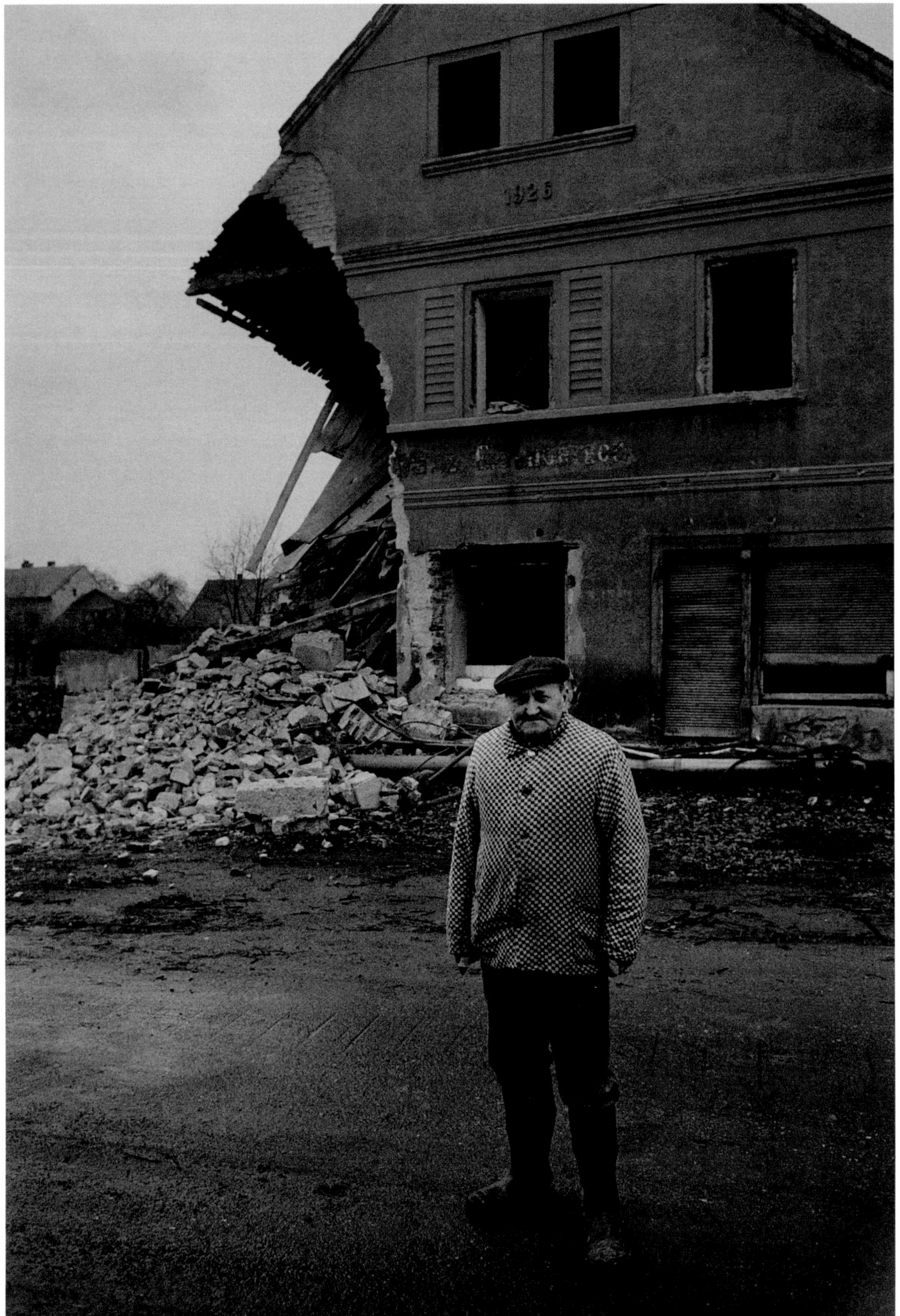

_ 030 ONE OF THE TOWN'S LAST REMAINING RESIDENTS
_ 031 DEMOLISHED TOWN MAKES WAY FOR COAL MINING

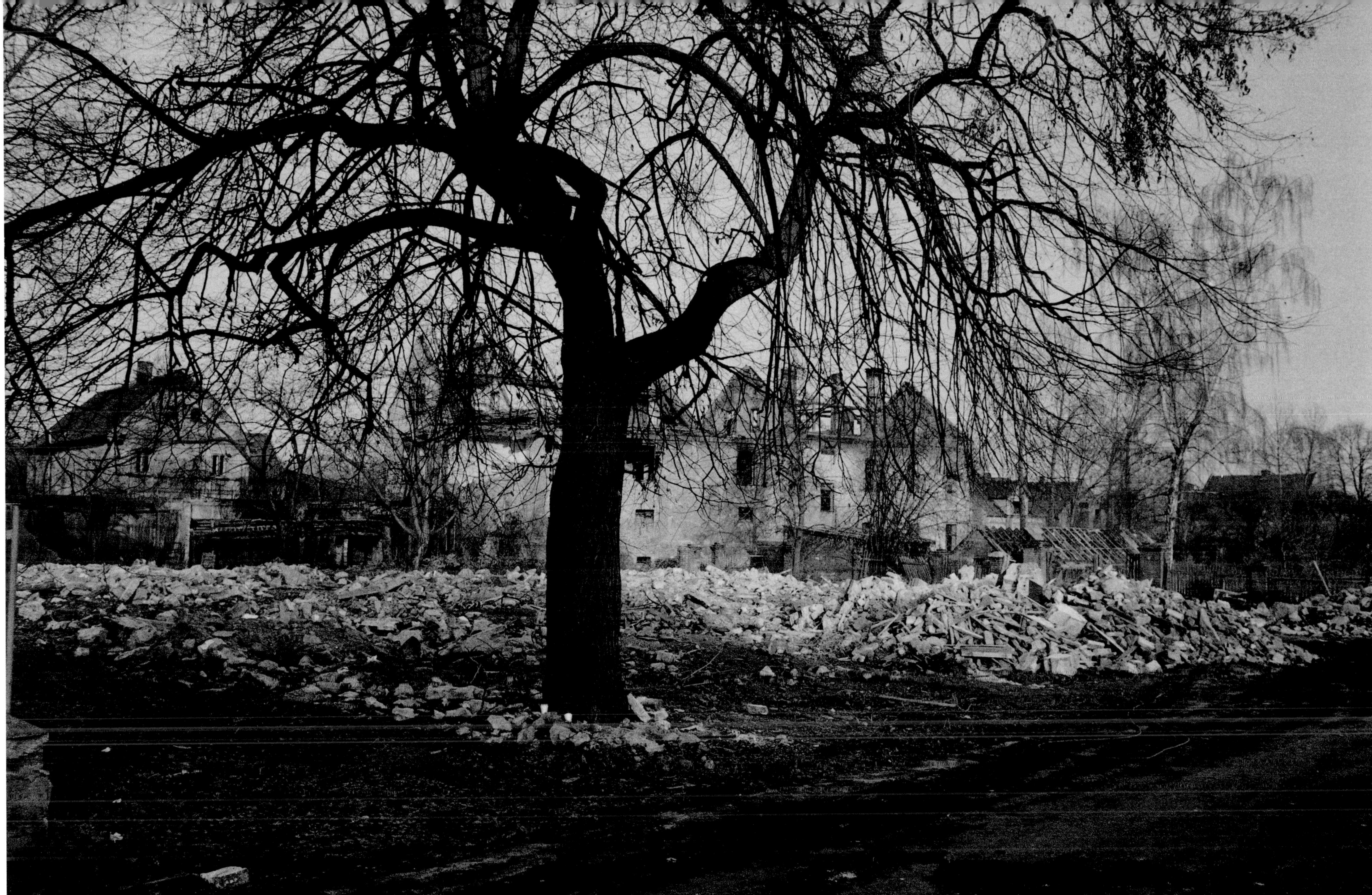

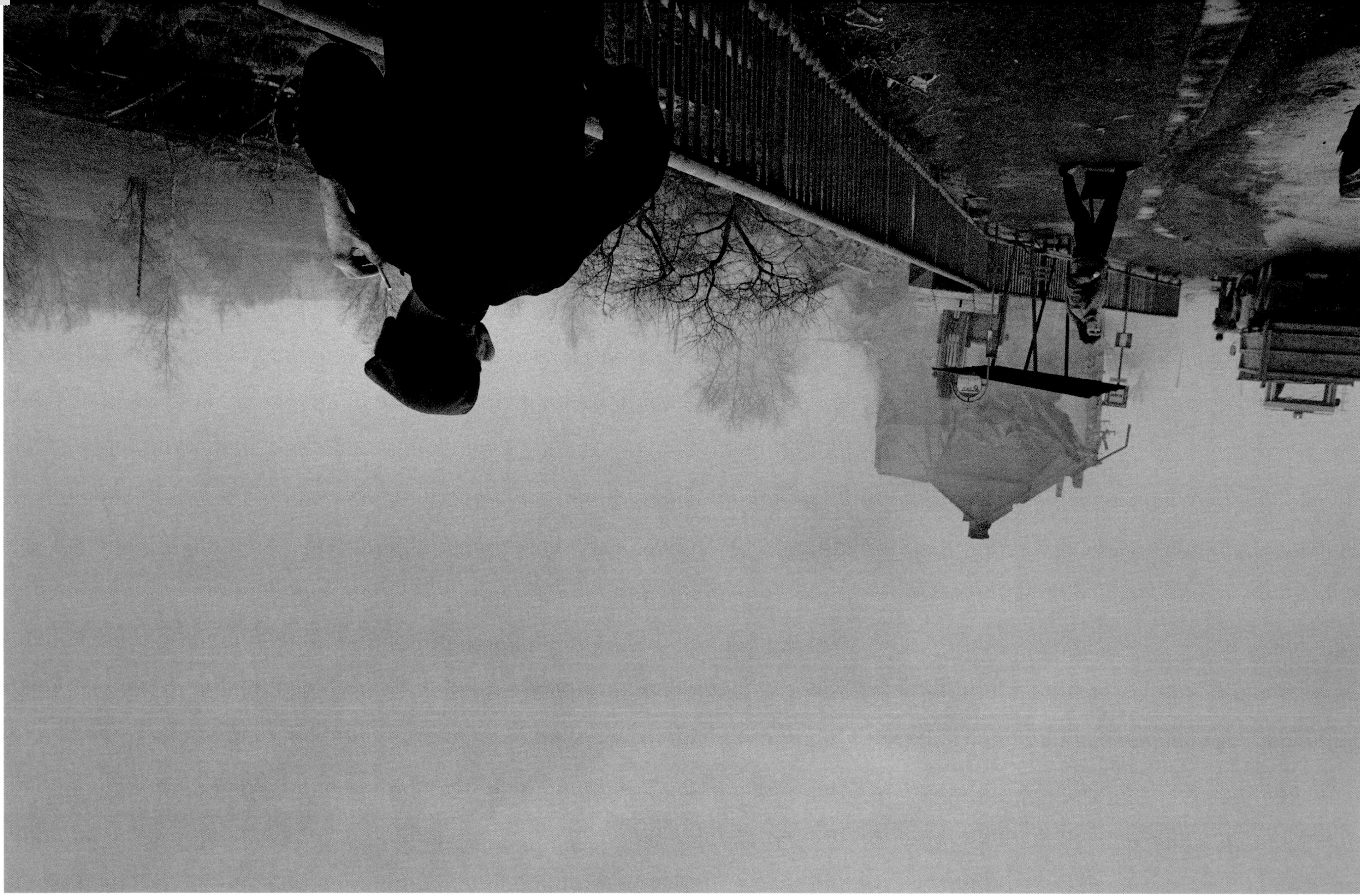

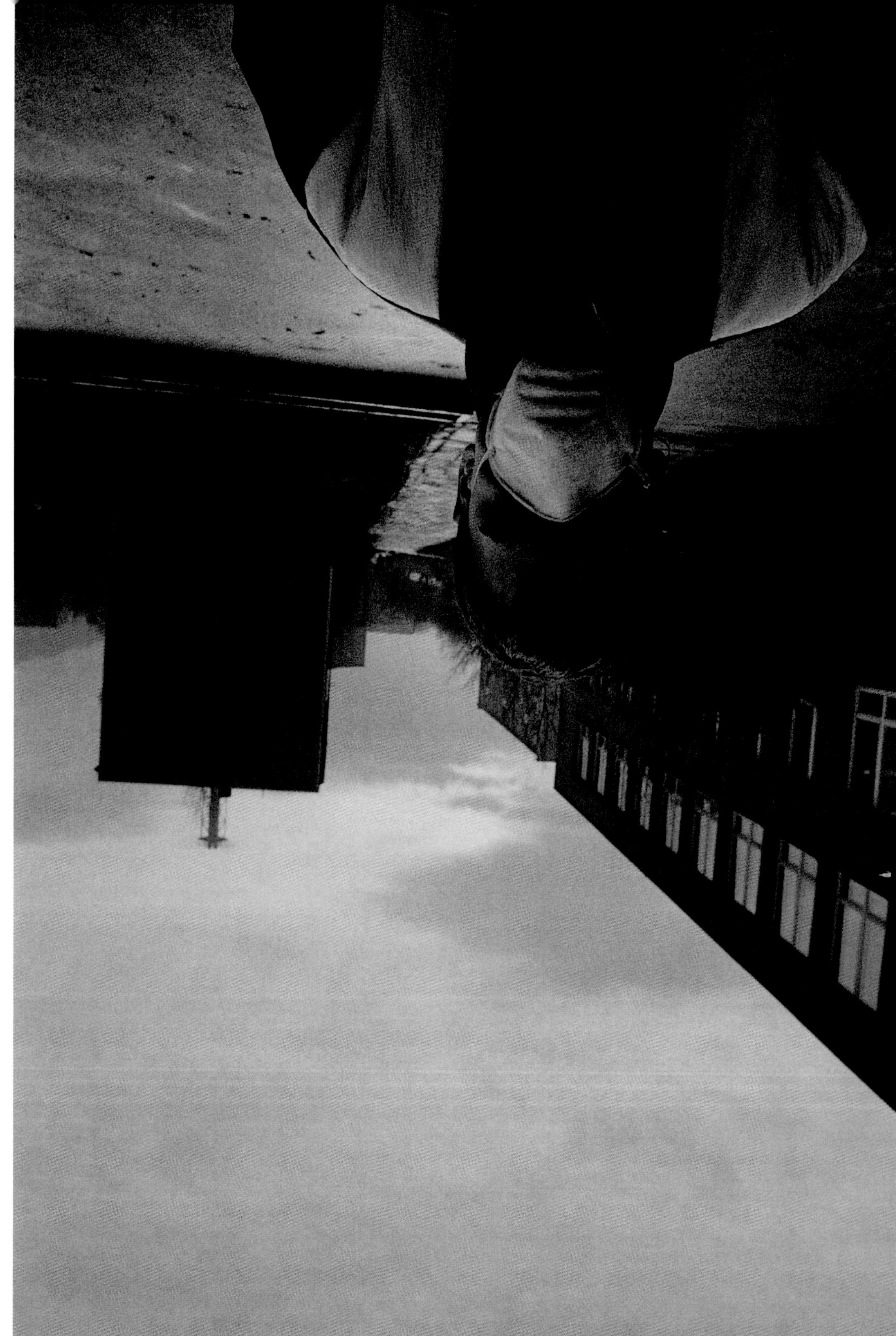

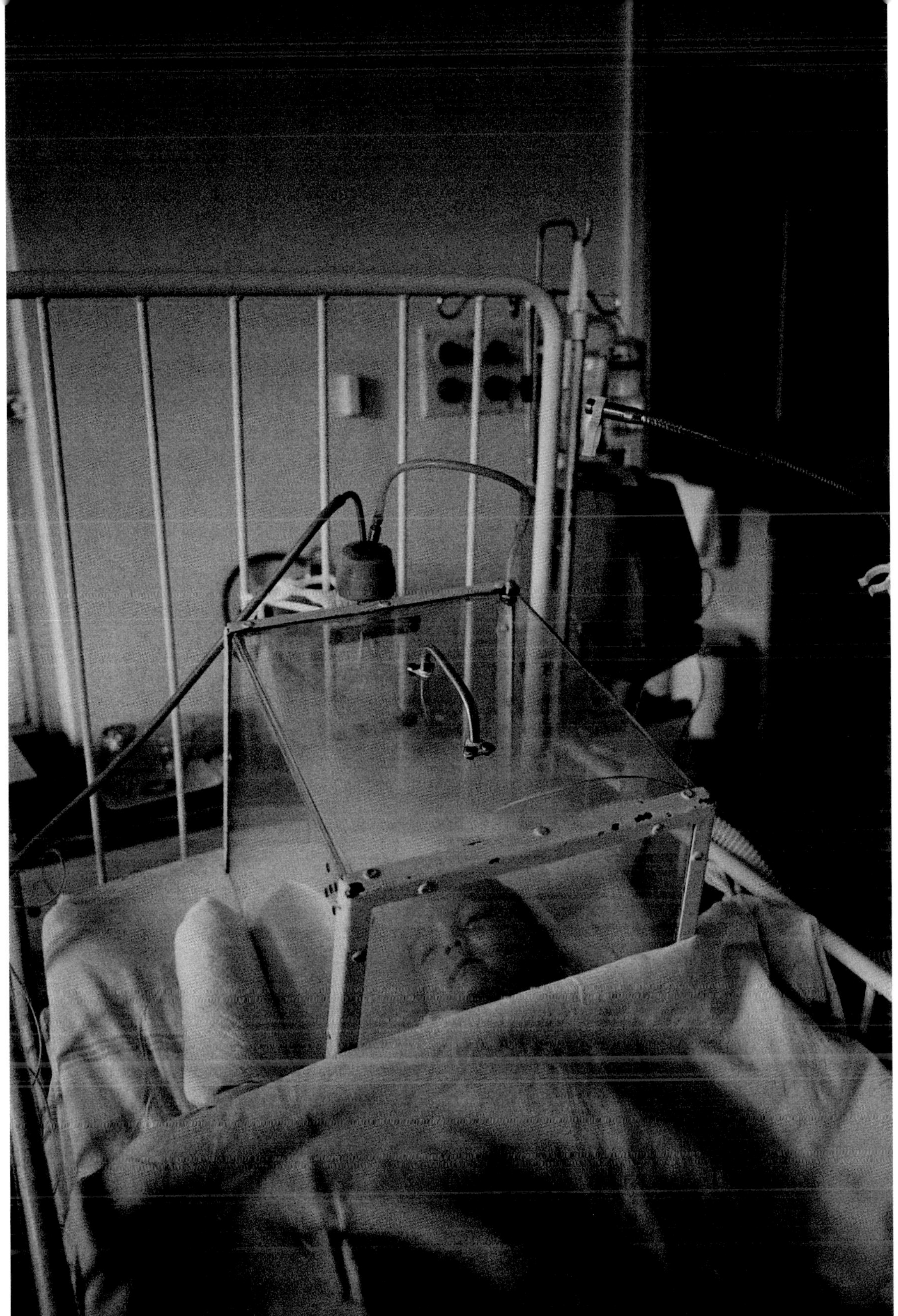

_ 034 STUDENT WEARS MASK AS PROTECTION FROM POLLUTION
_ 035 RESPIRATORY CLINIC

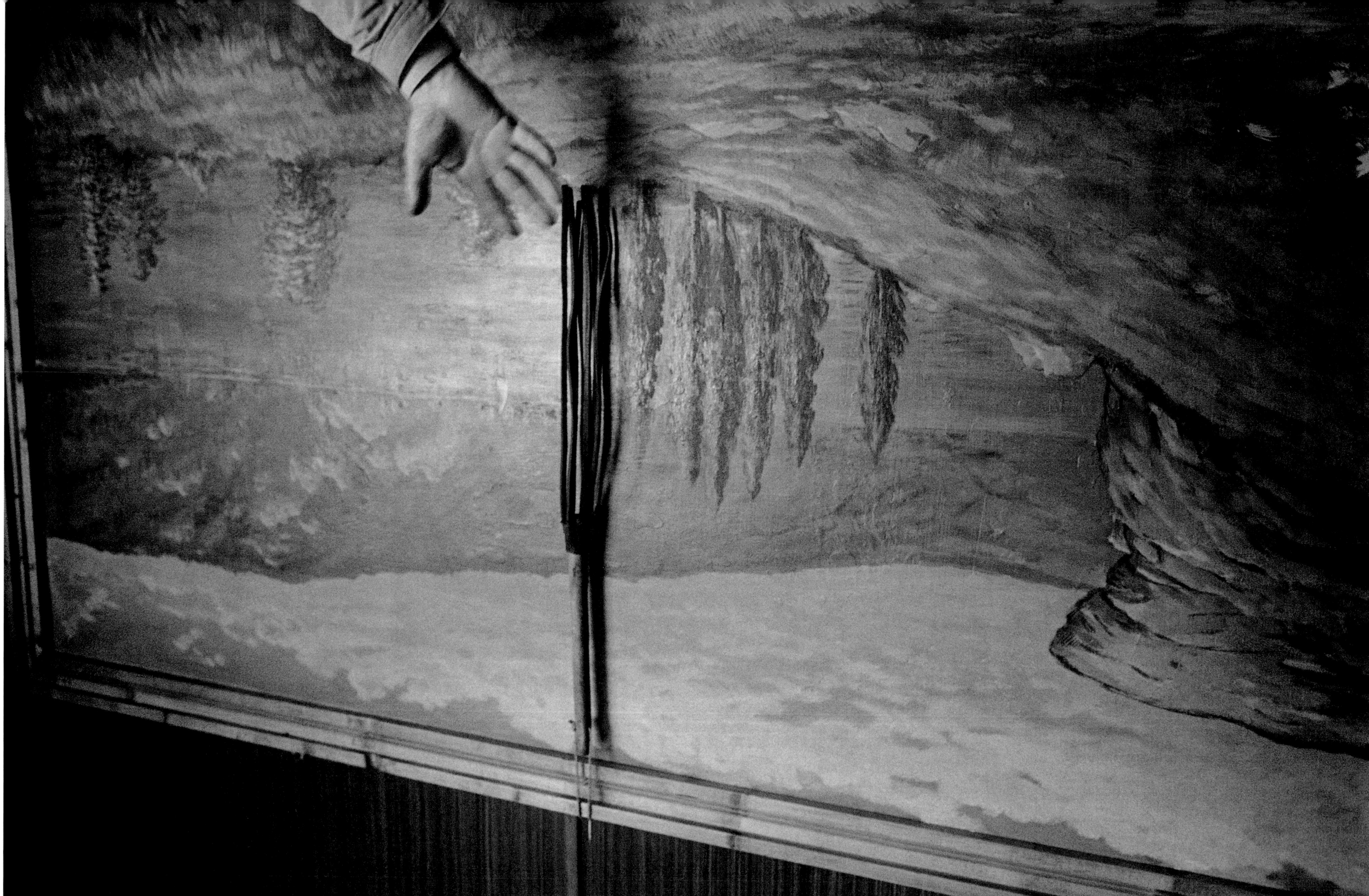

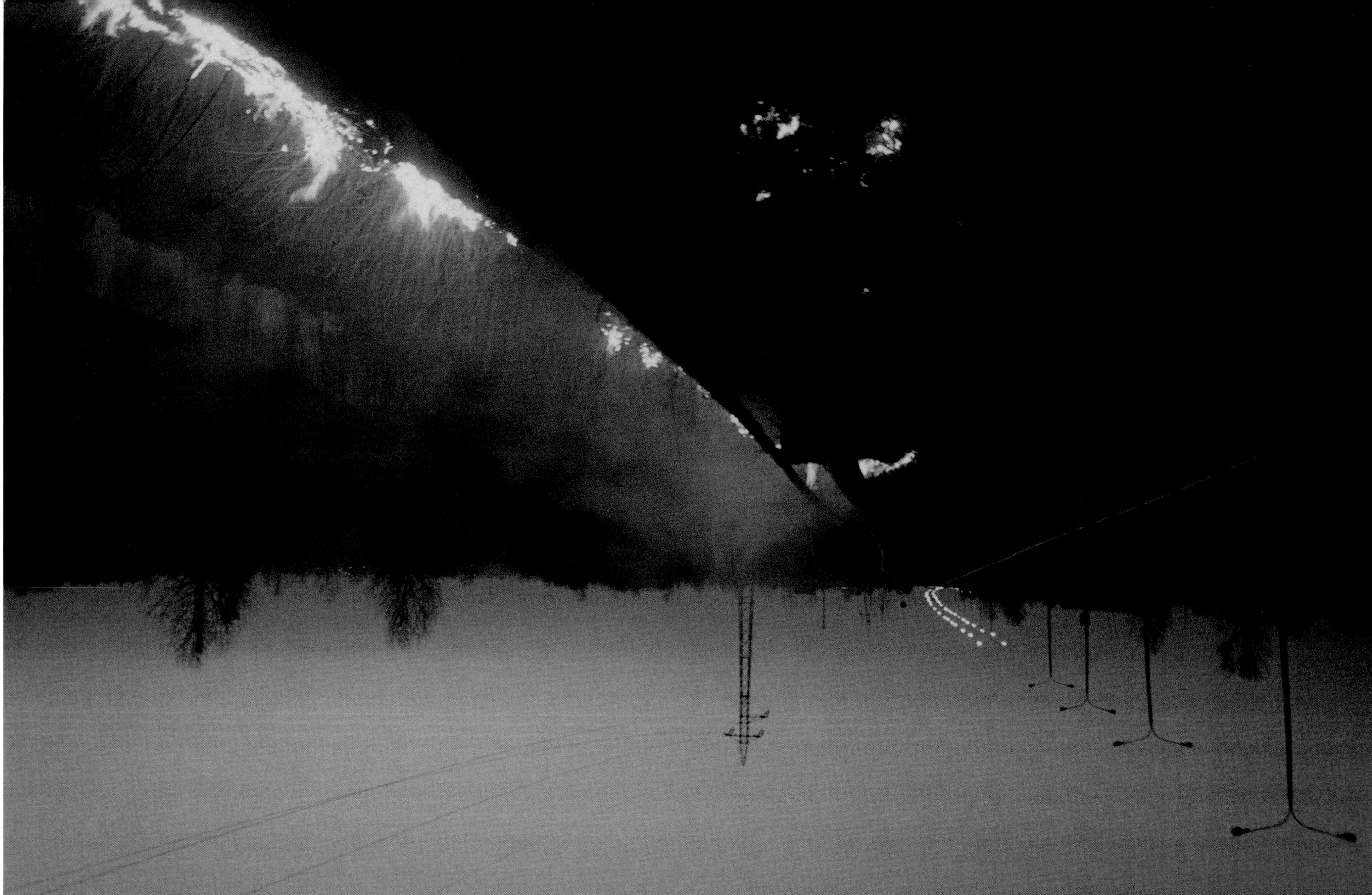

OVER A PERIOD OF TWELVE YEARS WHAT WAS ONCE THE MONTE CARLO OF THE RED SEA BECAME THE STALINGRAD OF THE MIDDLE EAST .

Muslims,
Christians and
Jews
did this.

Three sides, one God—Allah, The Almighty, HaShem.
When you ask,

WHO DESTROYED BEIRUT ?

it wasn't the SAM rockets fired by the P.L.O.,
the Syrian MIG's,
the Katyousha batteries used by the Druze militia or the phosphor bombs and implosion devices launched by the Israeli Defense Force.
No.
When you ask who did this,
tell them it was God.

GOD DID THIS.

GOD KILLED BEIRUT.

VENGEANCE IS MINE ; I WILL REPAY , SAYETH THE LORD .

Romans 21:19

beirut _ heaven sent

_ *Section*

III

HEAVEN SENT

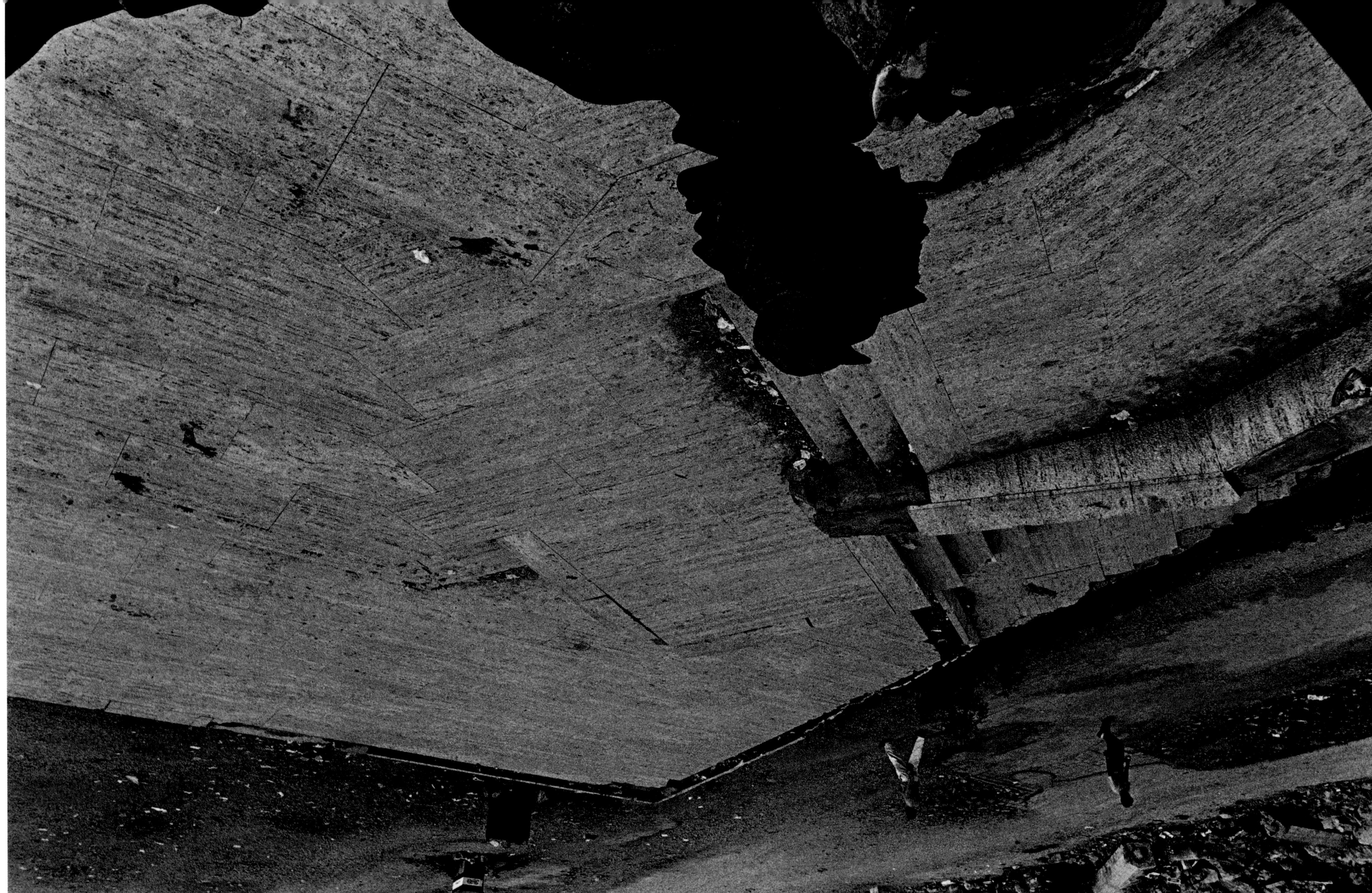

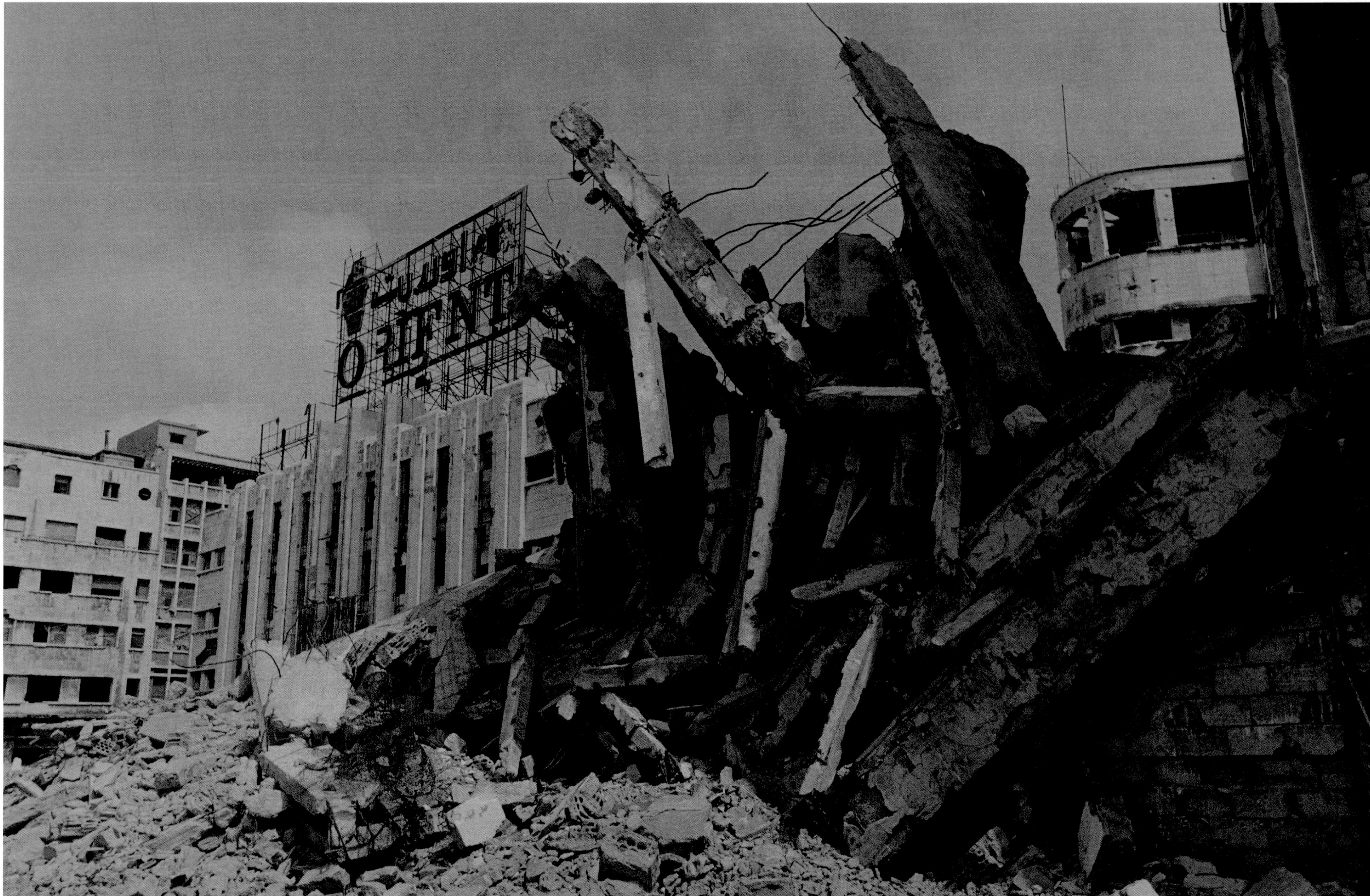
ORIENT

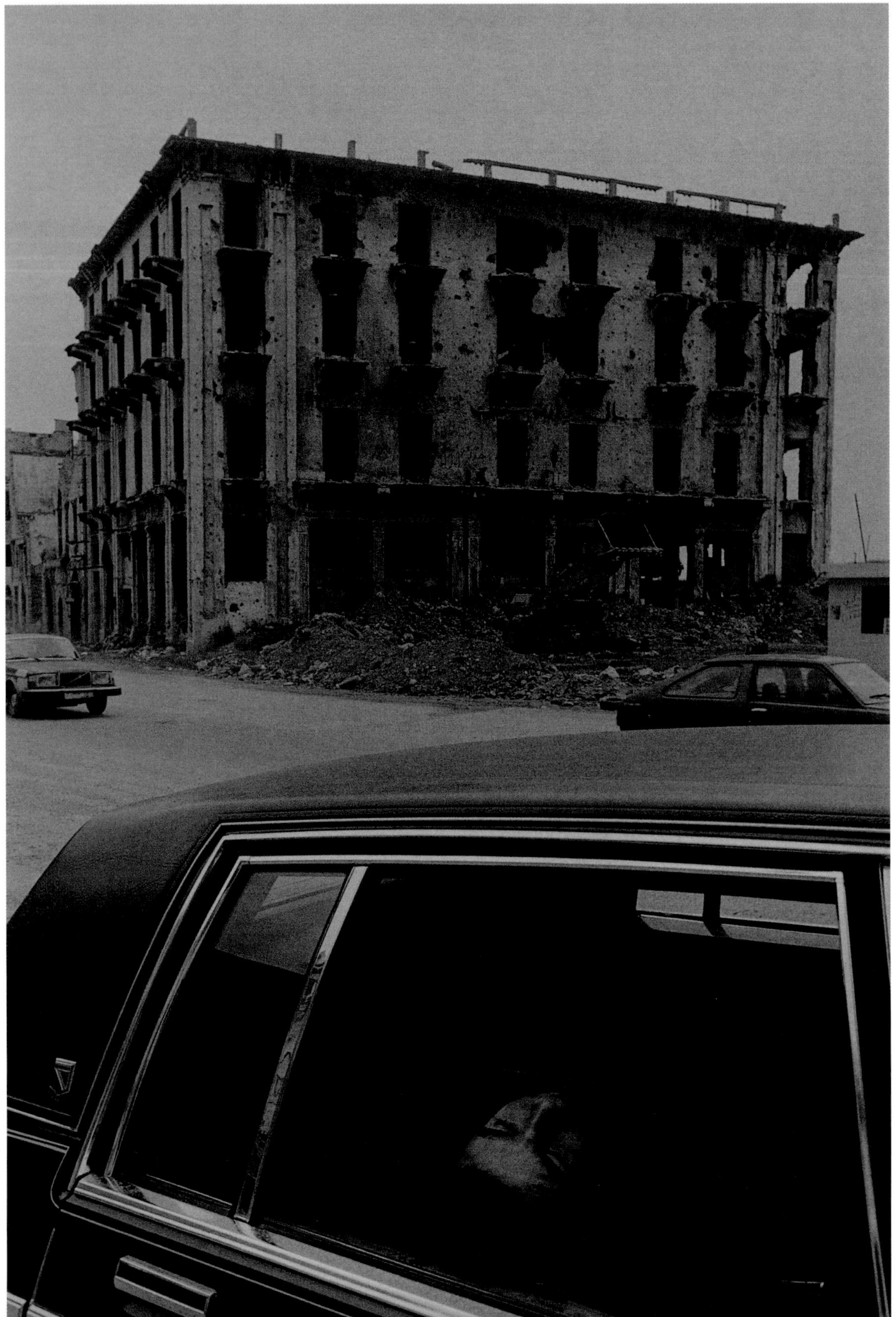

_046 SEASIDE
_047 DOWNTOWN

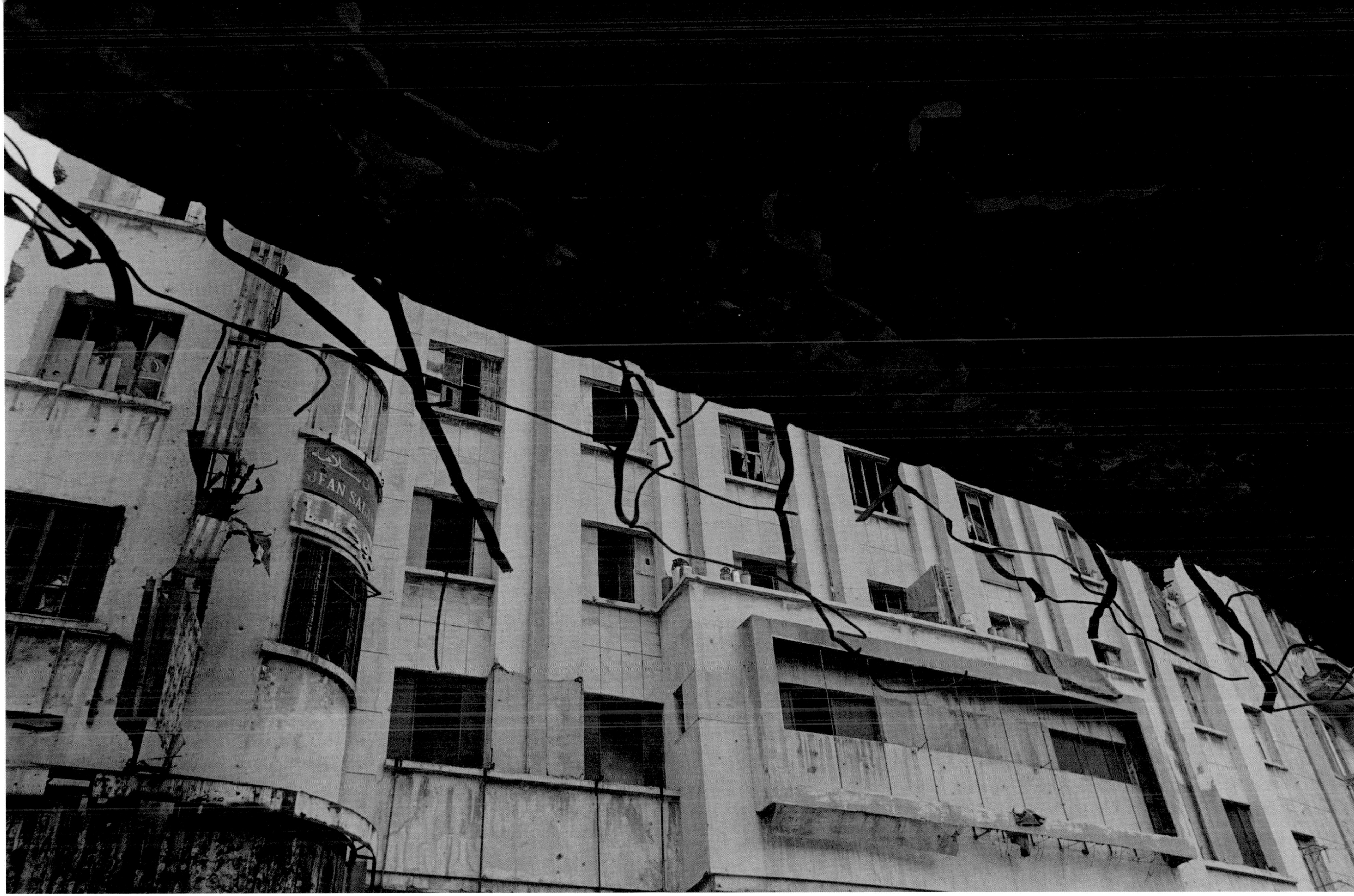

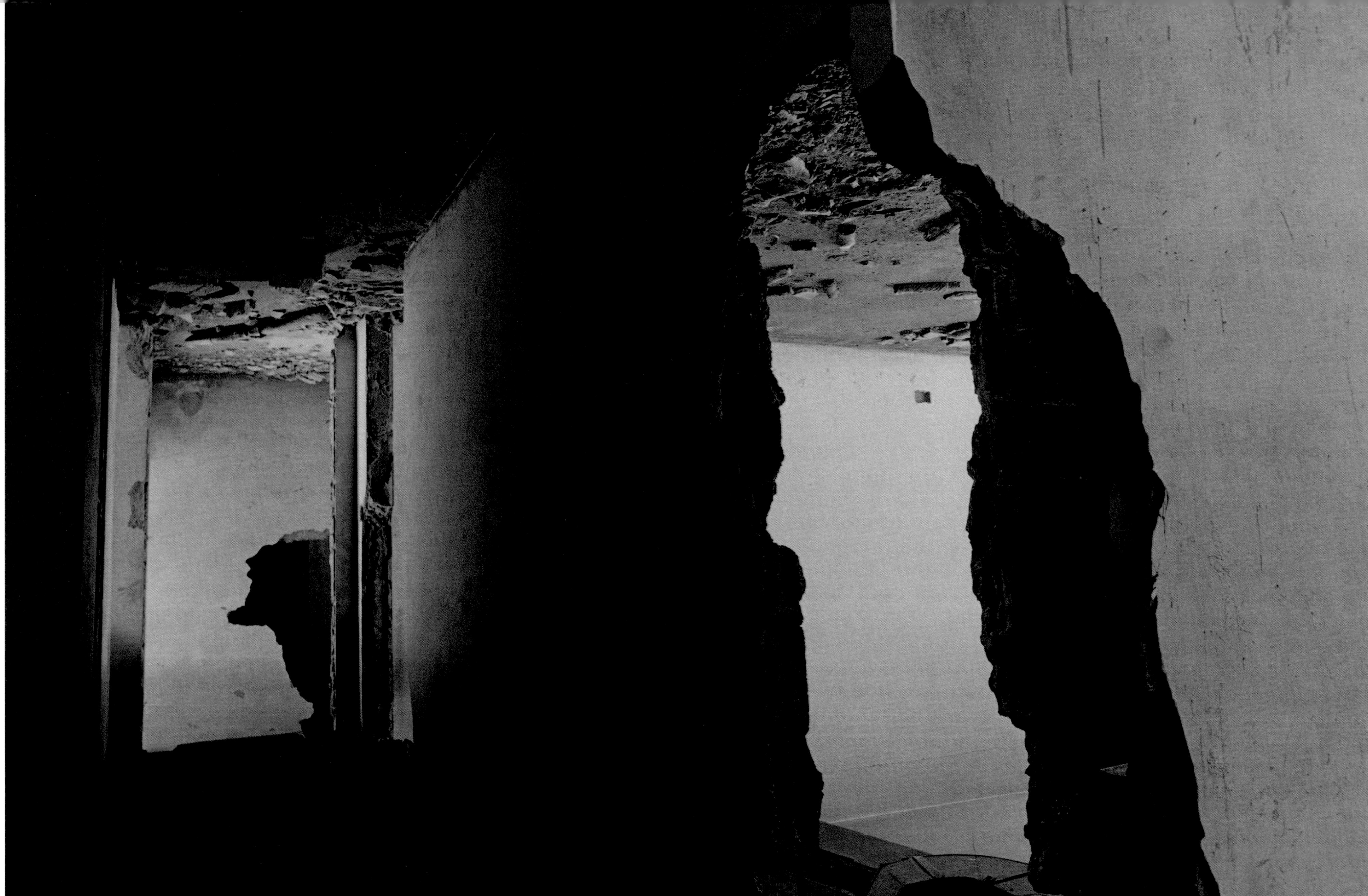

_ 051 TEA HOUSE AMONG THE RUINS
_ 052 SYRIAN CHECK POINT
_ 053 DRIVING DOWNTOWN

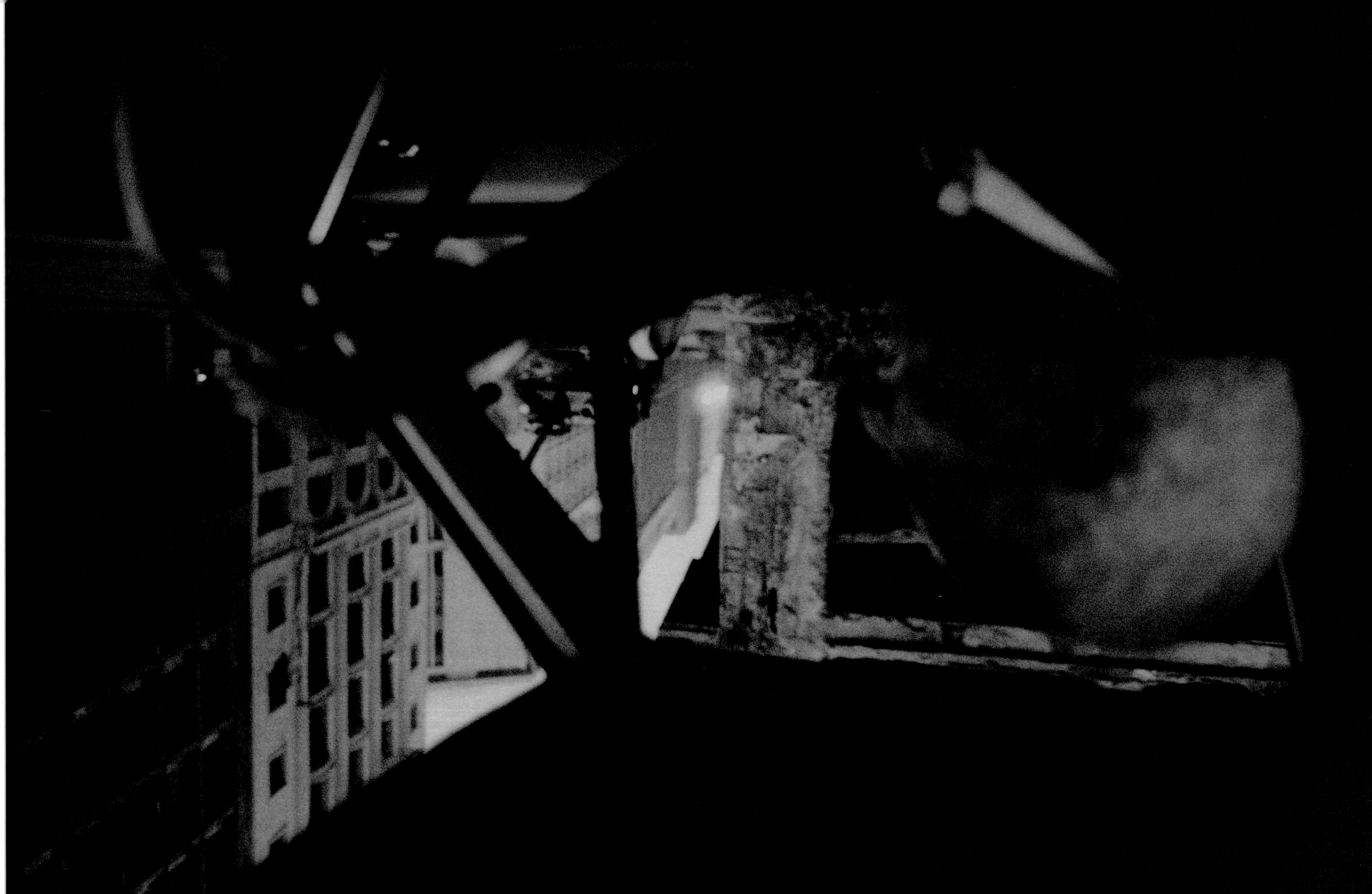

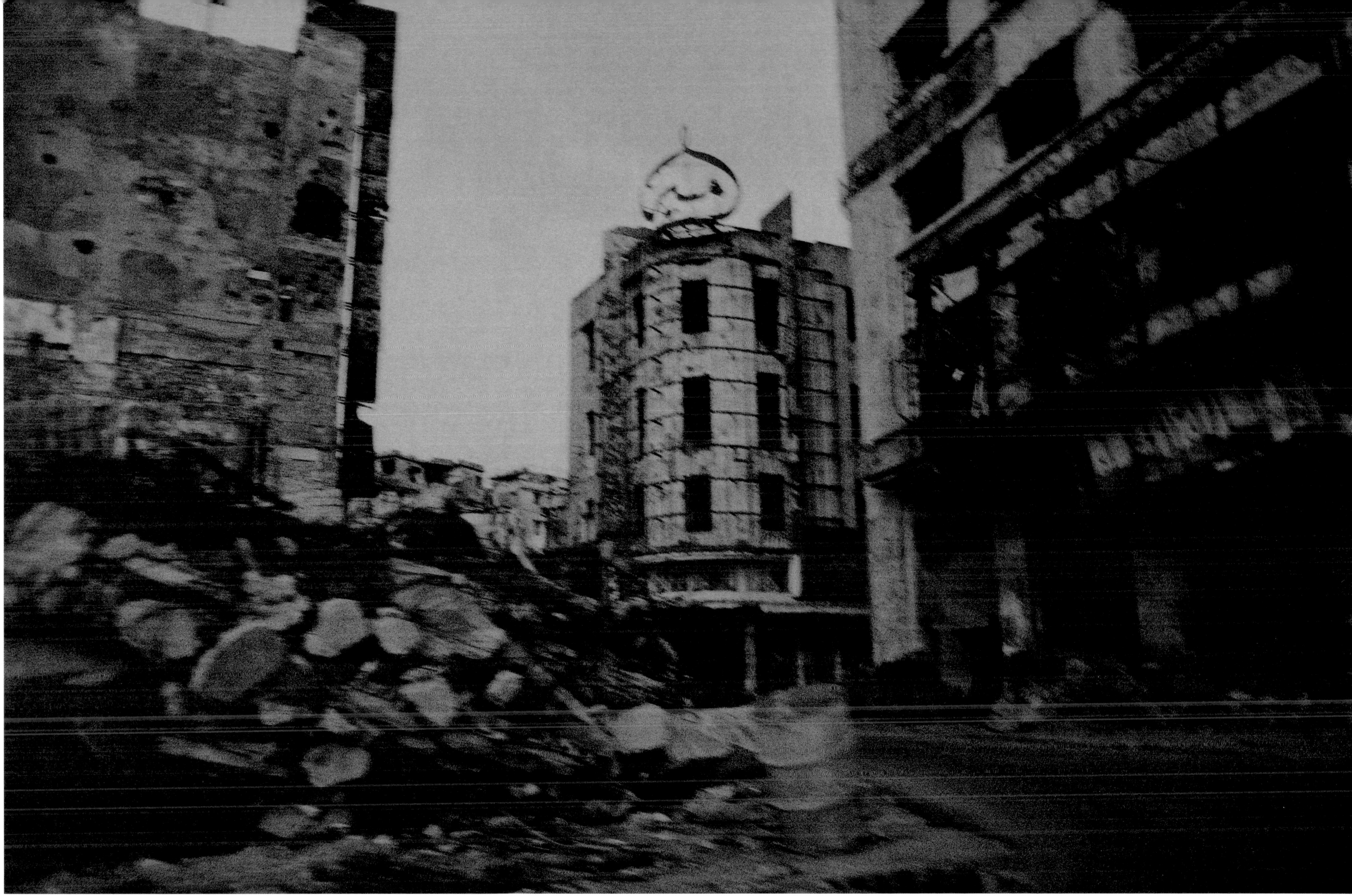

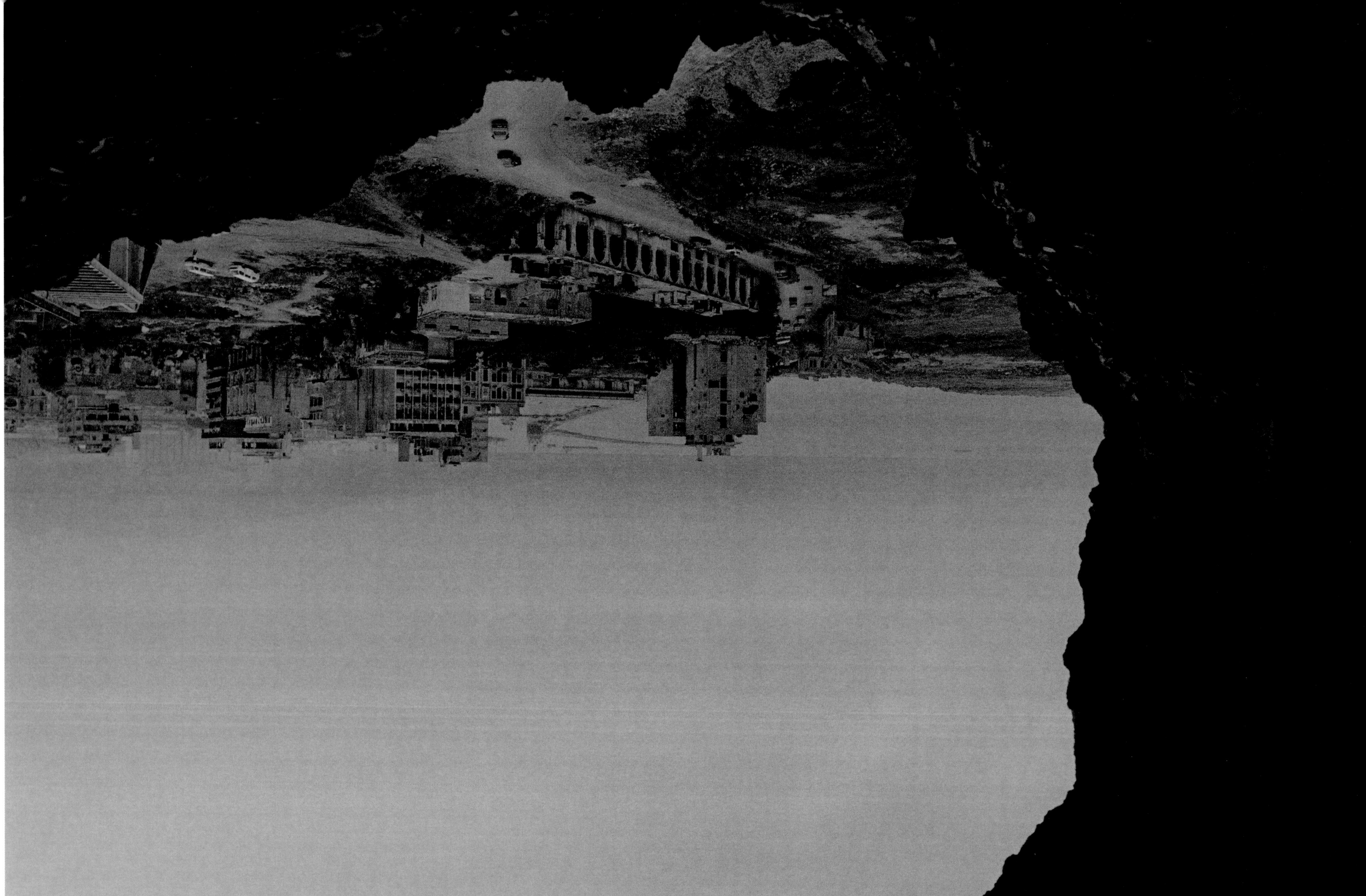

_ 055 HILTON HOTEL, SEASIDE VIEW

Tin once made Bolivia rich.

In fact, Bolivian tin made Simon I. Patino one of the 10 richest men on earth.
In 1904, he was crowned the KING OF TIN, and reigned supreme in the mountains southeast of La Paz.
There, he built the town of Llallagua for the workers who made him this metal monarch ... a place of luxury for them.

He gave them swimming pools,
a golf course,
theaters,
high schools,
hospitals;
he gave them commerce,
he gave them lives.

The mines of Llallagua like the legend of El Dorado made dreams come true for all the king's subjects.
The industry, like the ruler, became a legend.

The revolution of 1952 overturned all that.
The revolution that started a half century of corruption,
poor investment and the metal's overwhelming decline on the world markets.
left Llallagua's place of luxury and dreams in the dust.
The king had gone,
leaving his chattel and lands to the government bureaucrats,
industrialists and financiers who staked their claim in the country's revolving seat of power.
Patino's city in the mountains returned to rock as the mines closed down and ceased to be.

BUT WHAT OF THE WORKERS AND THEIR LIVES ?
WHAT OF THOSE WHO MADE TIN BOLIVIA'S LIFEBLOOD ?

The fortunate ones got out.
The others with nowhere to go and no money to leave remained behind.
They do so still today.
These industrial scavengers who toil in the slag heaps they once made are the *palliris*.
They are serfs to the shit that industry discarded.
They live in the rock and work with the rock and handle and see only rock.
They make just enough to linger in the wasteland and live only to sift through the past's debris that the present has forgotten out of hand.
They are barely in existence, barely able to account for life.

THE PALLIRIS ARE PHANTOMS .

For the *jucos*, the independent miners, stealing ore from the poorly run, state owned pits is the trap they cannot escape.
It is their only means.
For them there is opportunity to earn more than the scavenger palliris that rummage through the rubble.
Stealing makes them more,
if more describes something greater than subsistence.
In the end, the *jucos* are no different.
What they earn from their theft, they pay double in jail time.

THE JUCOS ARE PHANTOMS , TOO .

walks,
or crawls,
blending into the grey that is all around.
And they wait.
They wait for better times to bring another king,
another Tin King to come again and give his tin men back their lives.

IV :

bolivia _ tin men

_ *Section*

IV

TIN MEN

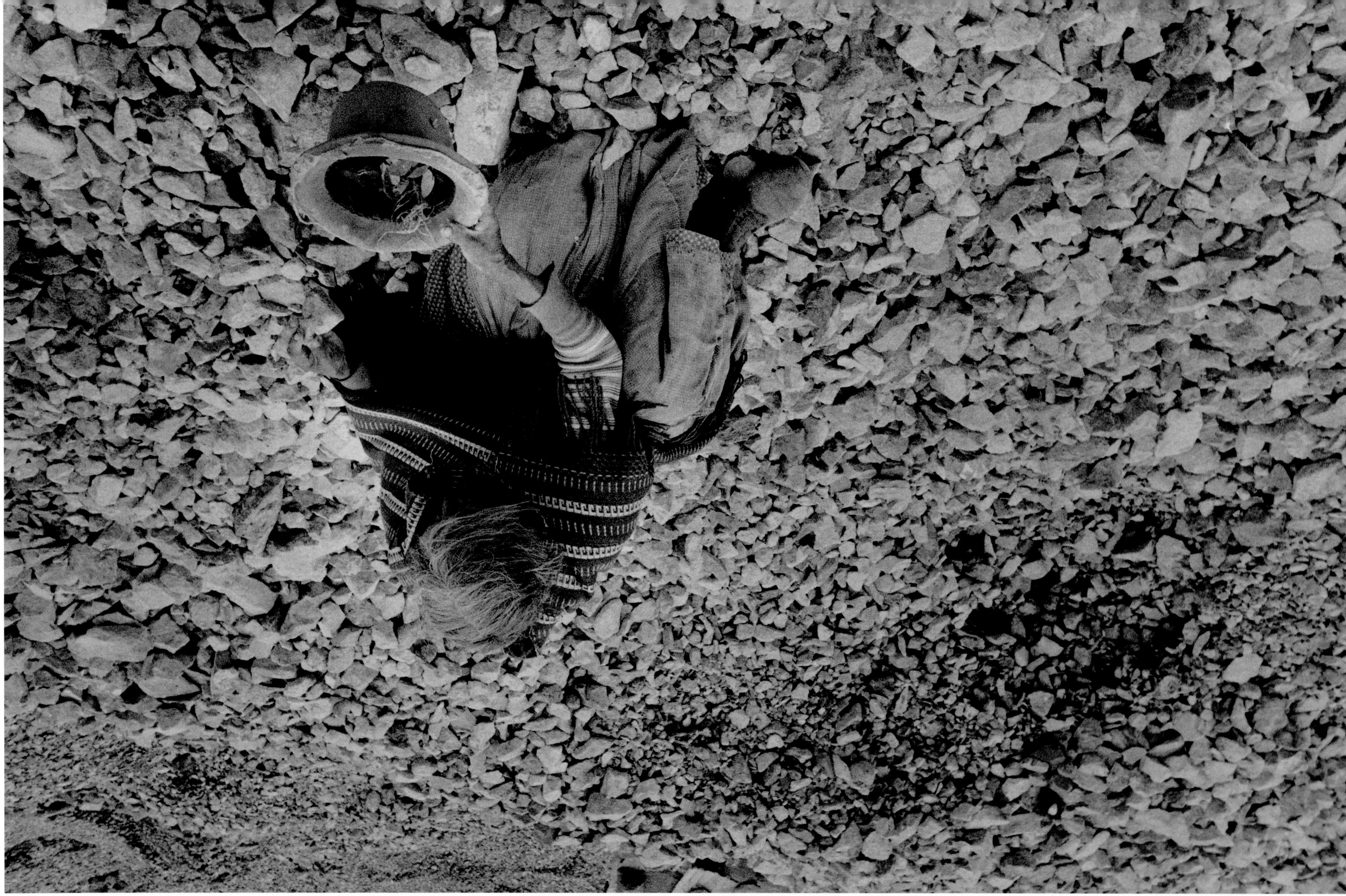

_ 059 SCAVENGING THROUGH THE MINE TAILINGS
_ 060 ARRESTED MINERAL THIEVES
_ 061 ARRESTED MINERAL THIEVES

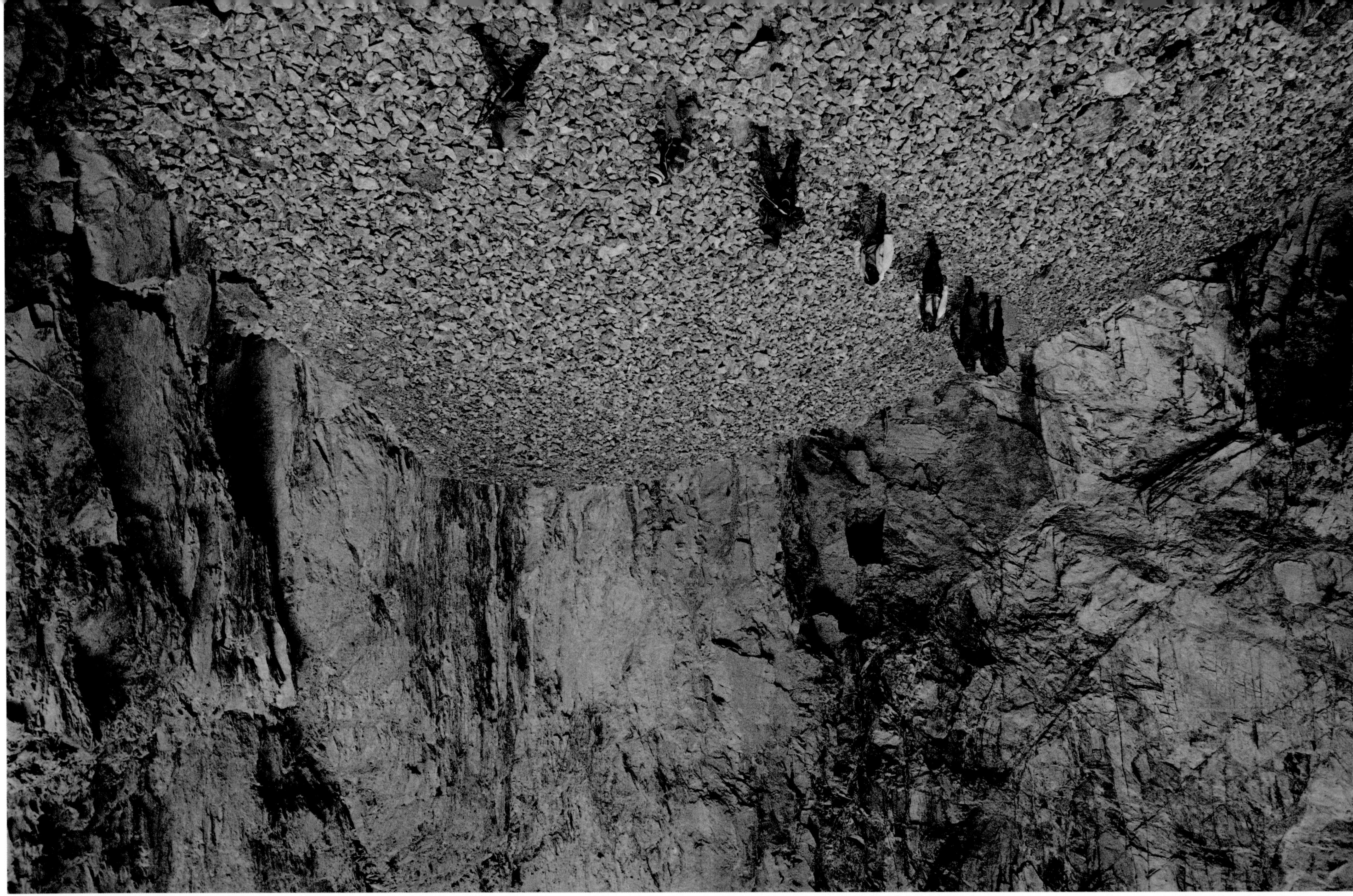

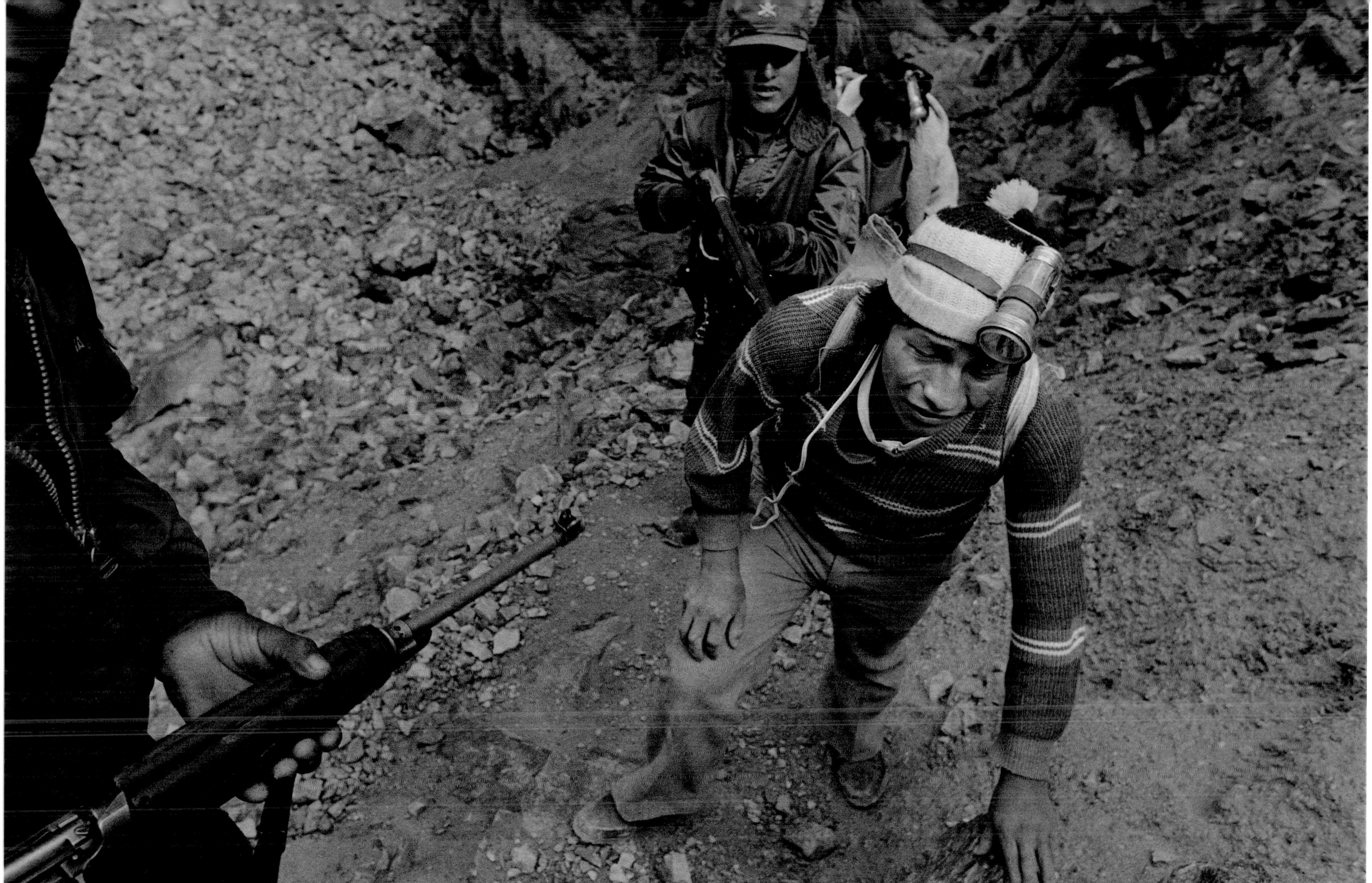

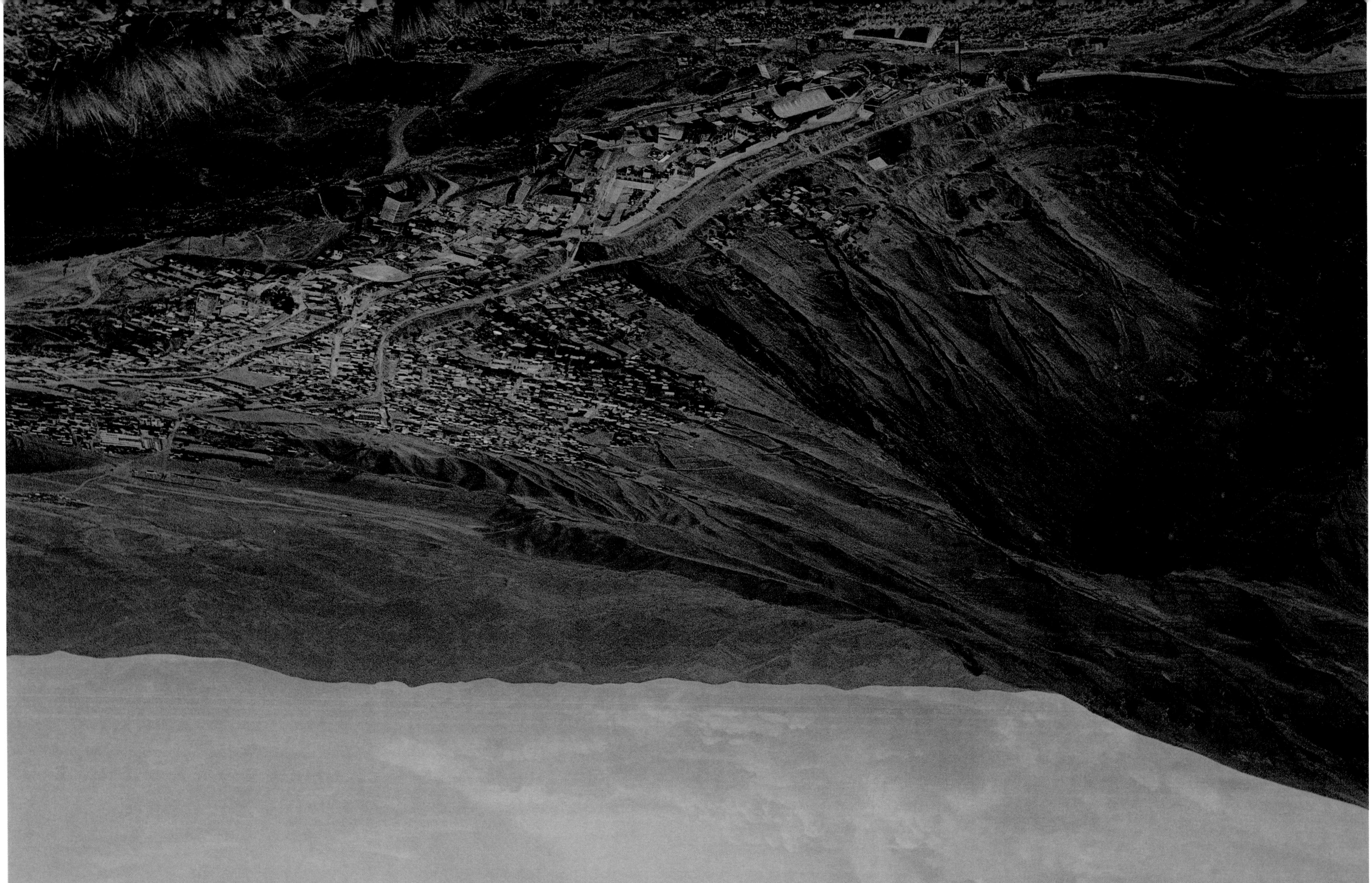

_ 063 TIN MINING TOWN
_ 064 ABANDONED TIN MINE
_ 065 PALLIRIS, WITH A BAG OF COCA LEAVES

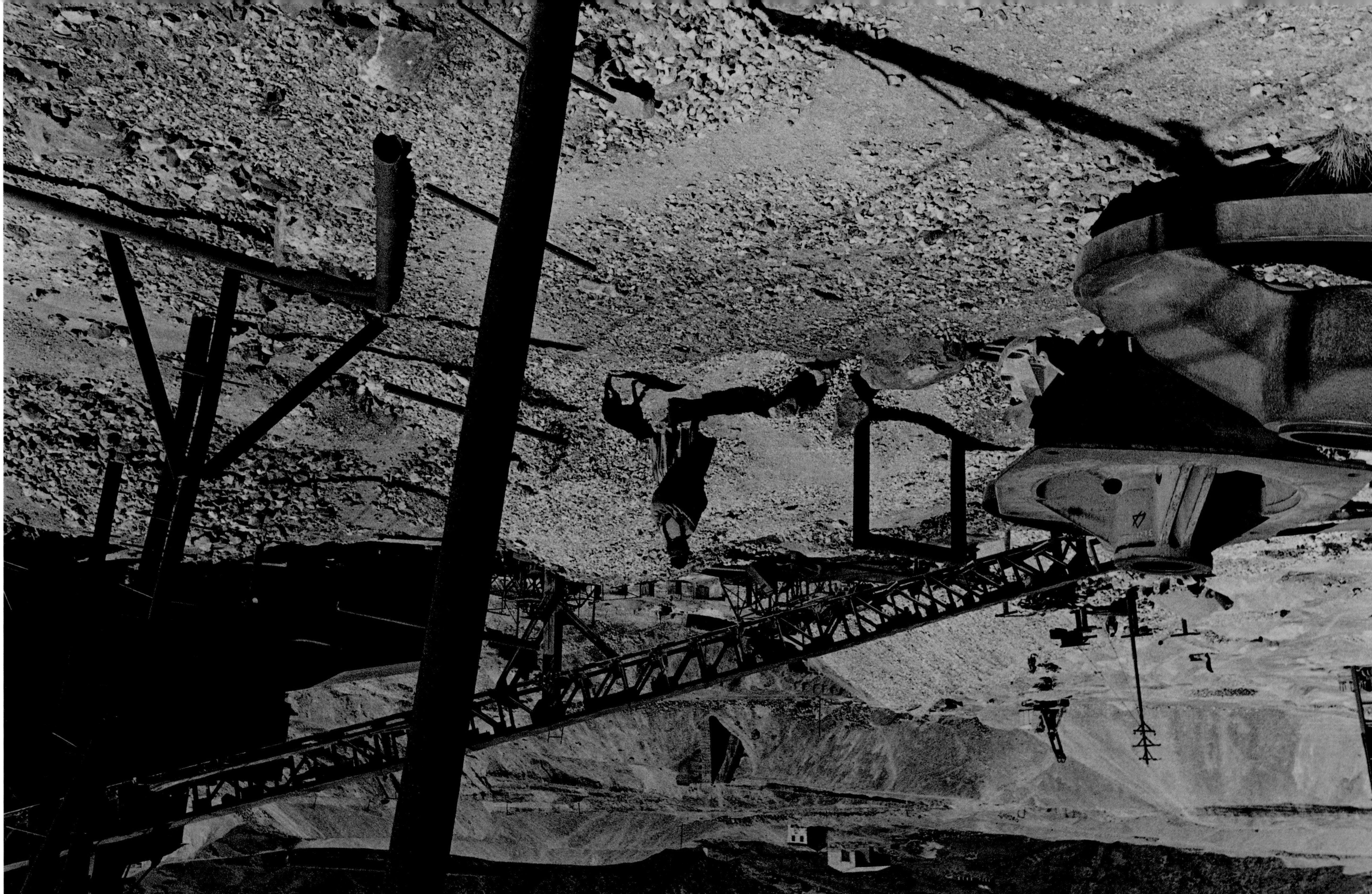

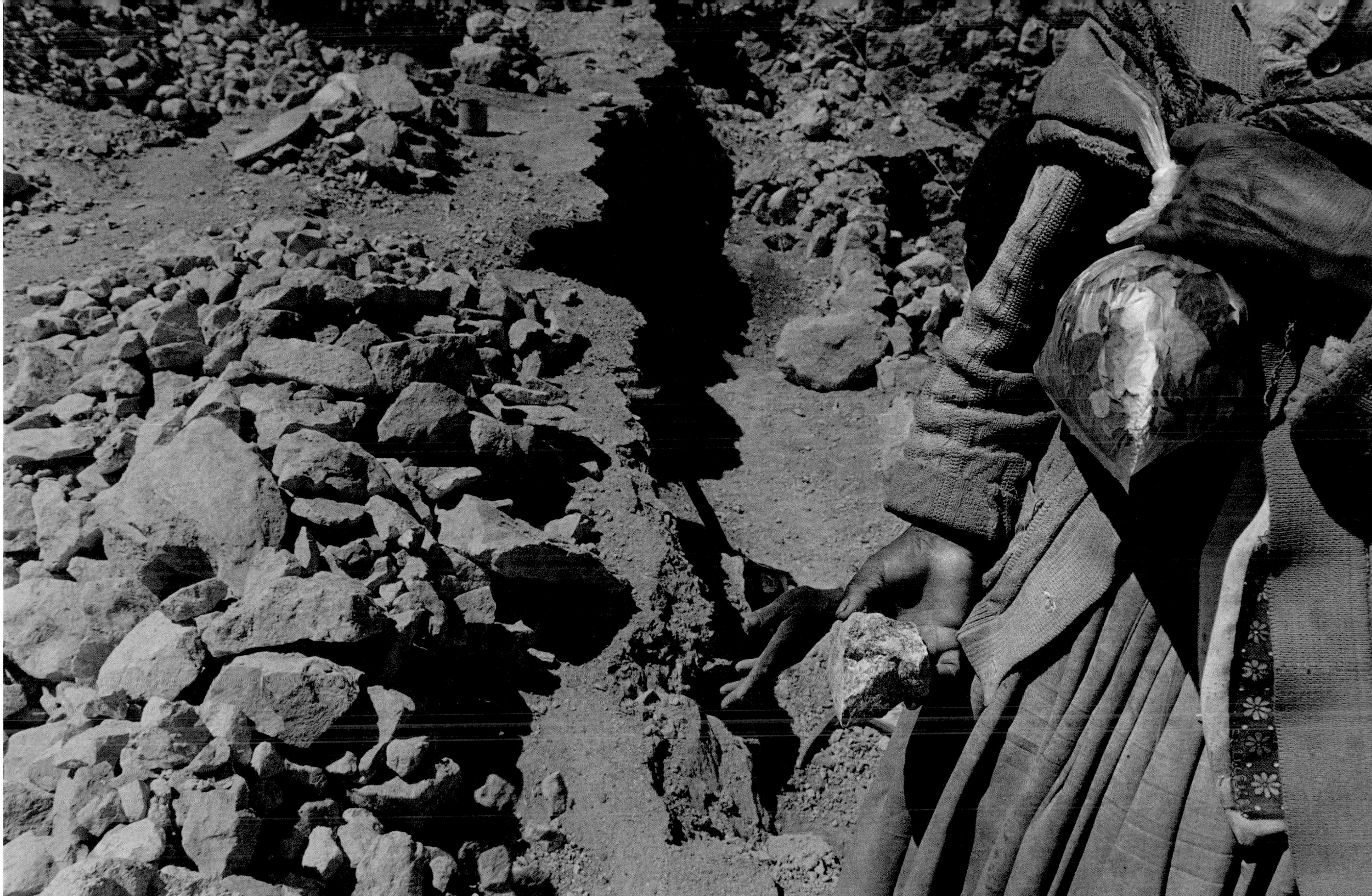

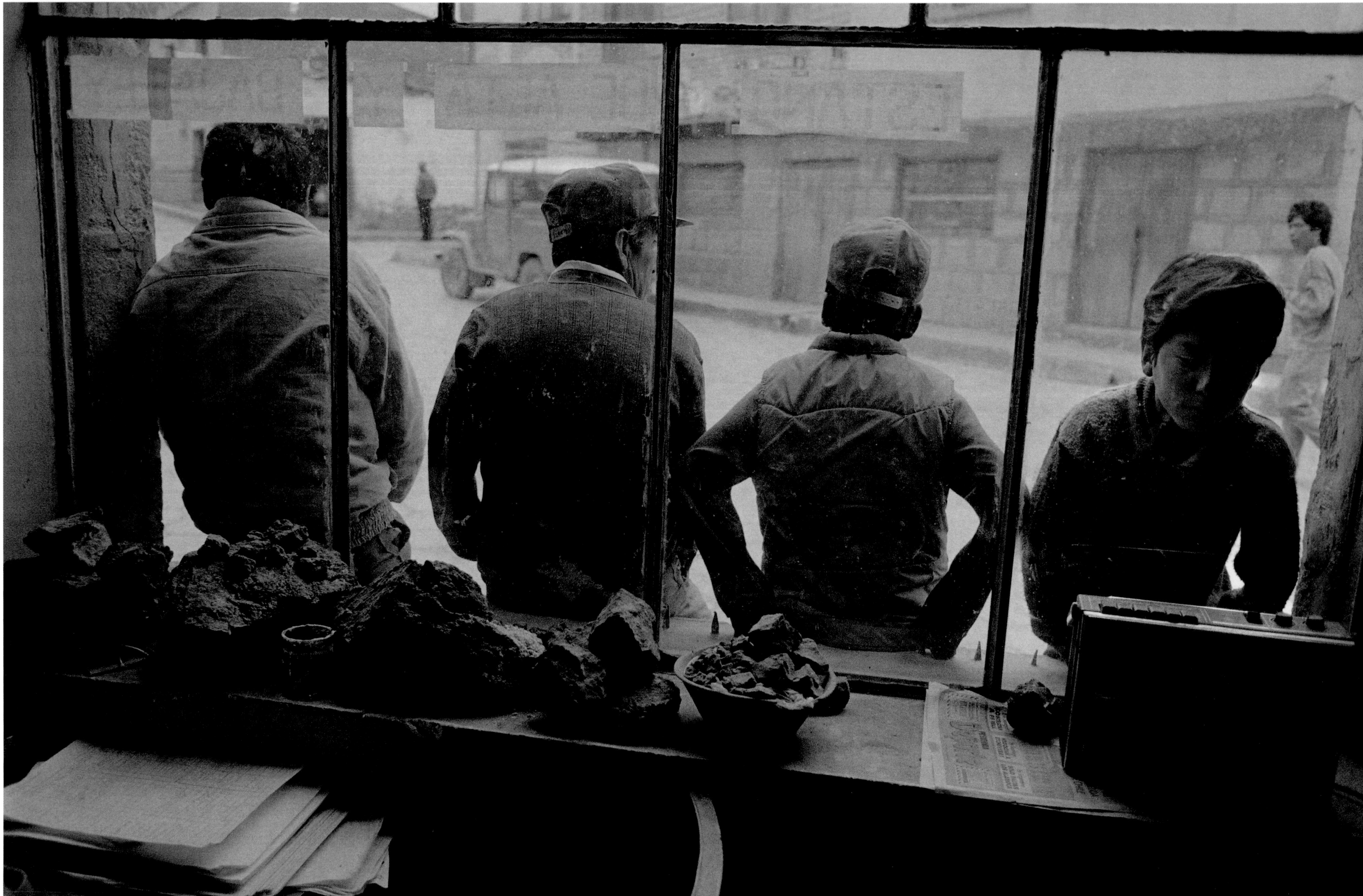

_ 067 MINERS WAIT FOR THEIR MONEY

“ A passenger called me over … there was a problem on carousel four.
It was the afternoon …
that’s when most West African flights get in.
As I got there I saw was blood streaked across everything.
I mean everything was smeared in blood,
all over the bags, on the equipment …
everything.
We located the suitcase … got it off the carousel,
and opened it up.

INSIDE, WHAT LAY THERE CHOPPED INTO EIGHT PIECES WAS AN ANTELOPE.

The animal hadn’t been cleaned,
gutted, or anything.
It looked like it had been slaughtered on the runway and dumped in a bag. ”

A BRITISH CUSTOMS OFFICIAL, HEATHROW AIRPORT, UNITED KINGDOM.

Widespread poaching and the activities of logging companies in the Congo Basin are estimated to produce just under 5 million tons of bushmeat (off the bone) per year. Wildlife cannot survive such slaughter. At its rate the mountains of meat being made and traded will leave the region’s wildlife extinct within 15 years.

International Primate Protection League

congo _ tropical take-away

_ *Section*

V

TROPICAL TAKE-AWAY

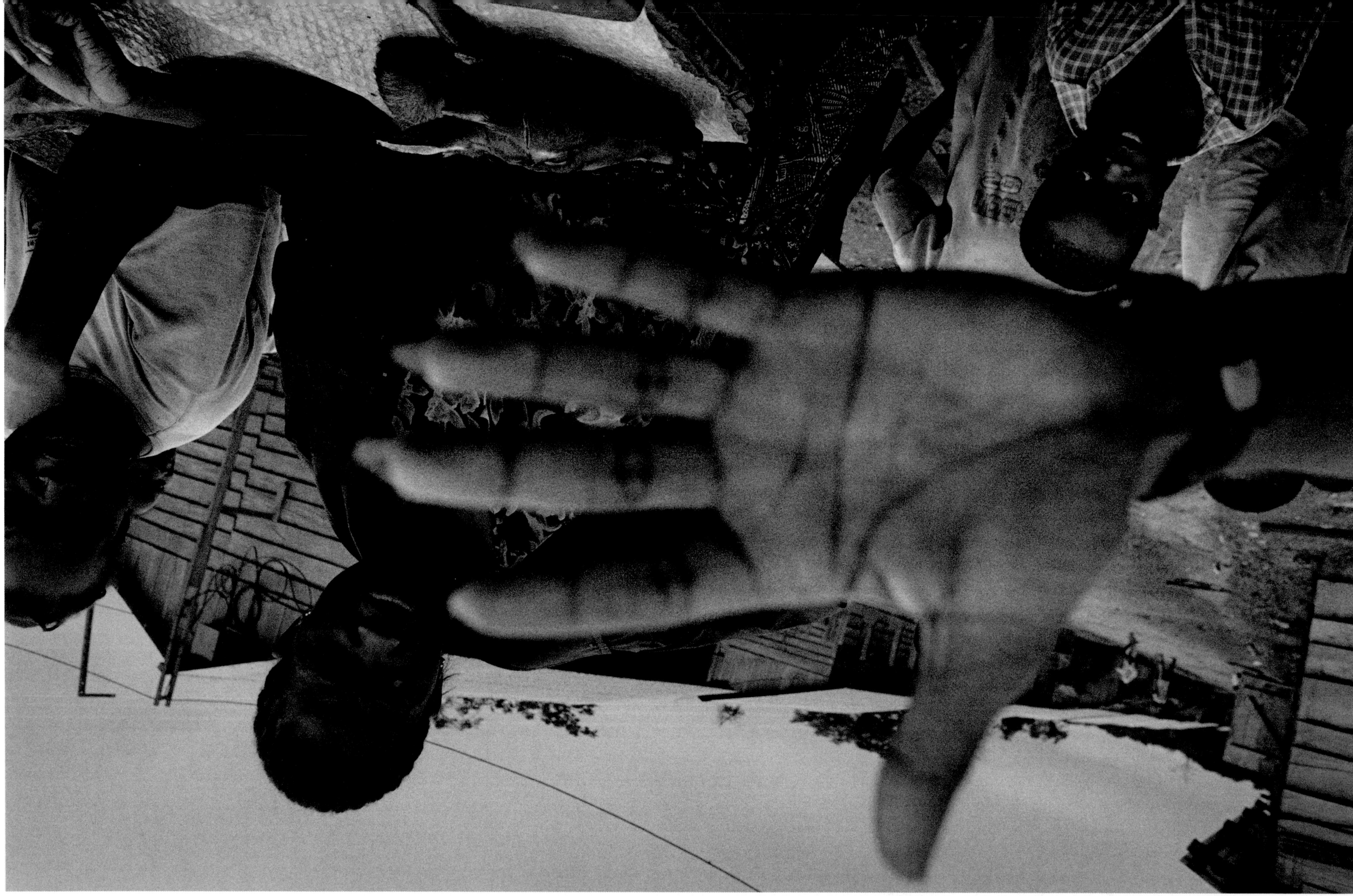

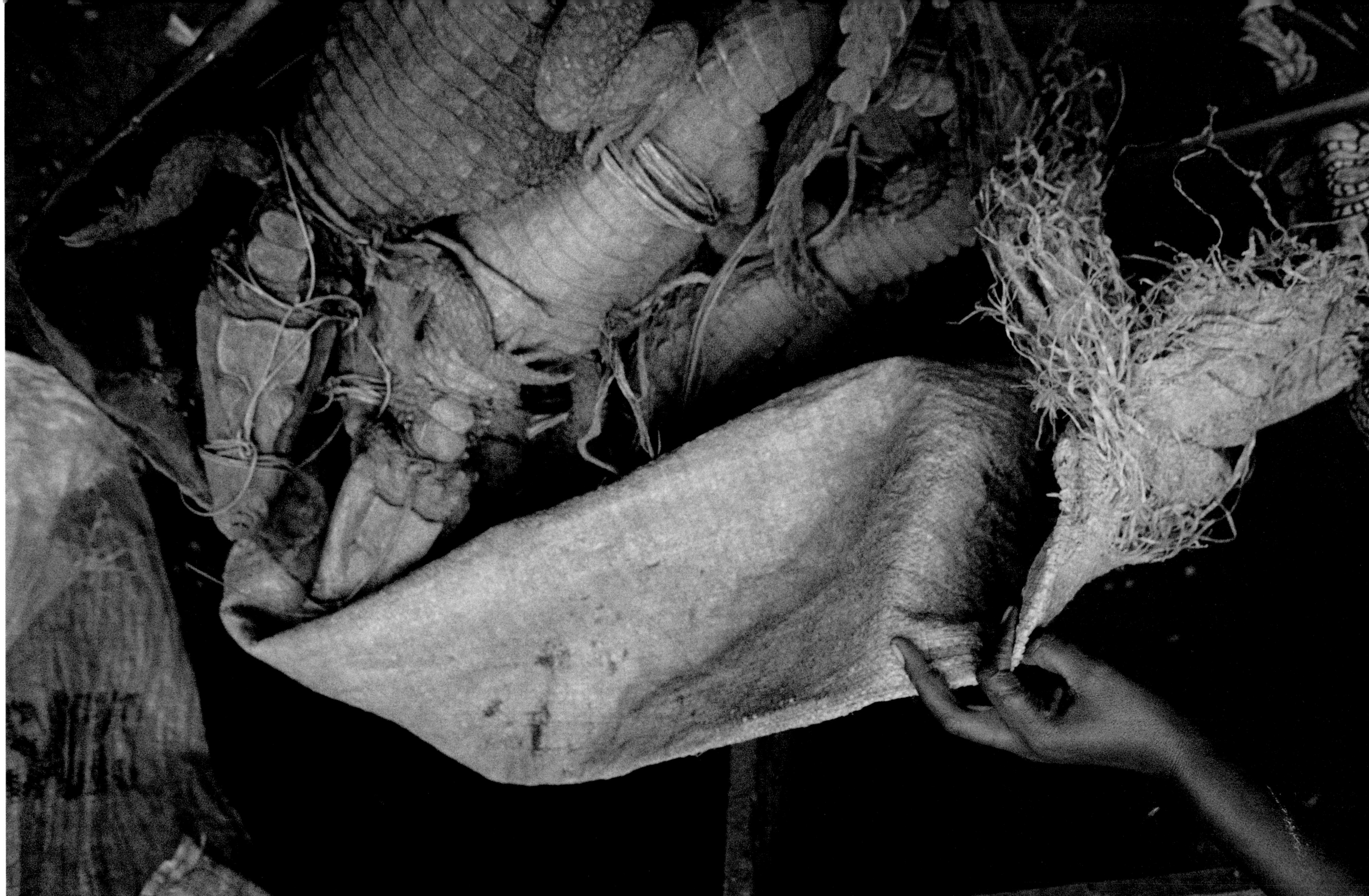

_ 075 ECO GUARDS ON PATROL
_ 076 ECO GUARDS WITH POACHERS' CATCH
_ 077 POACHERS' CONFISCATED GUNS, NETS AND IVORY

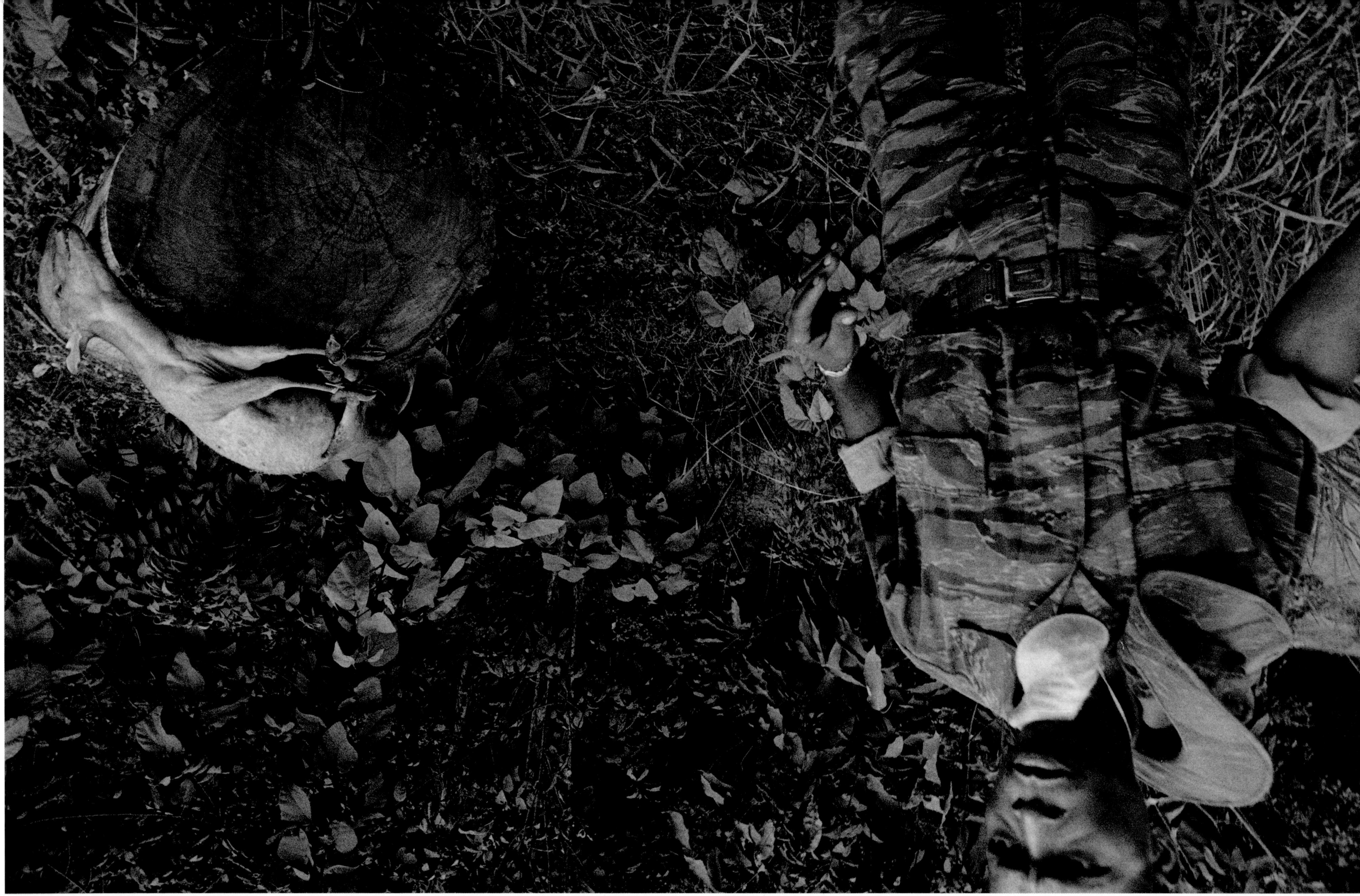

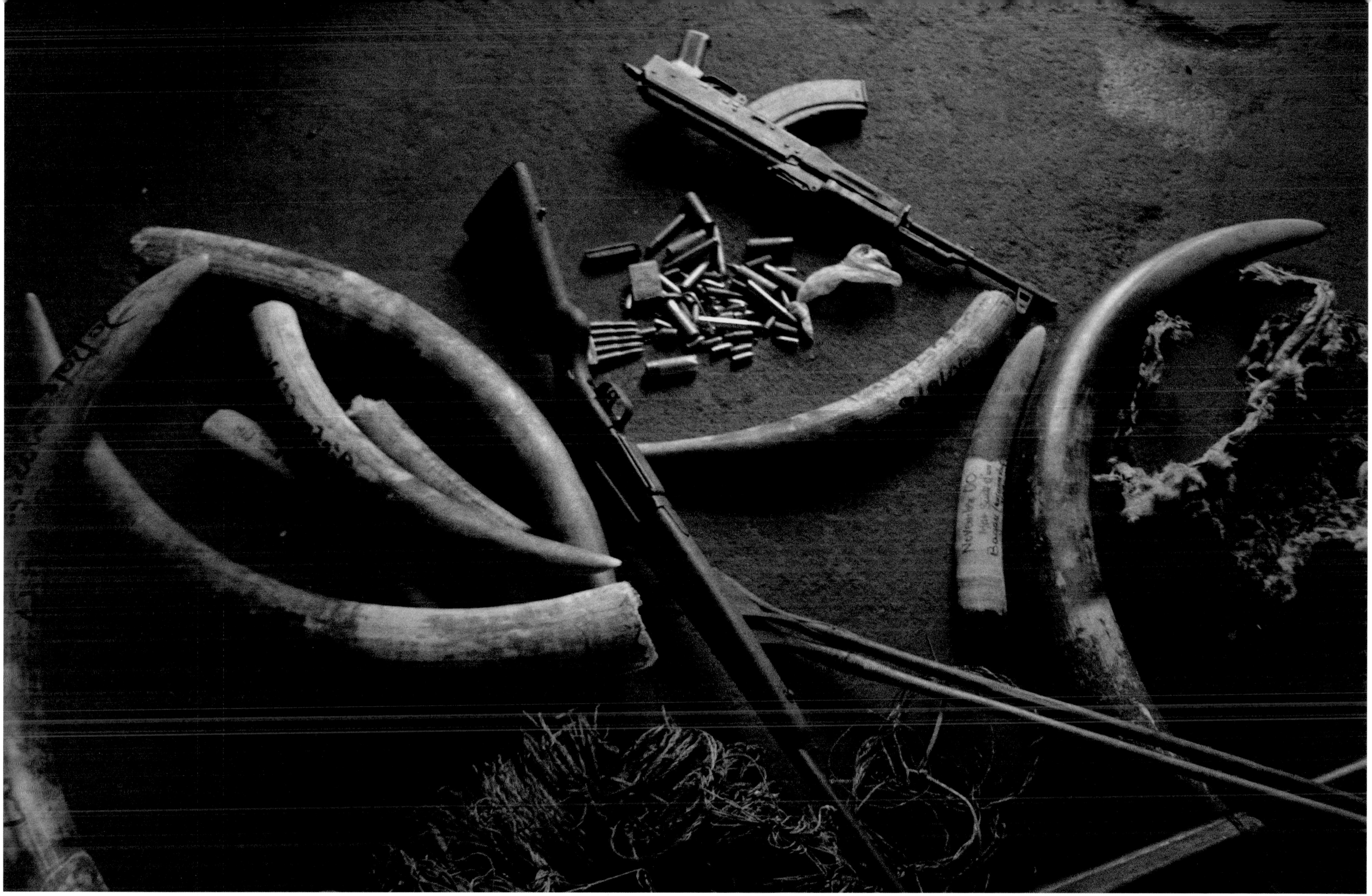

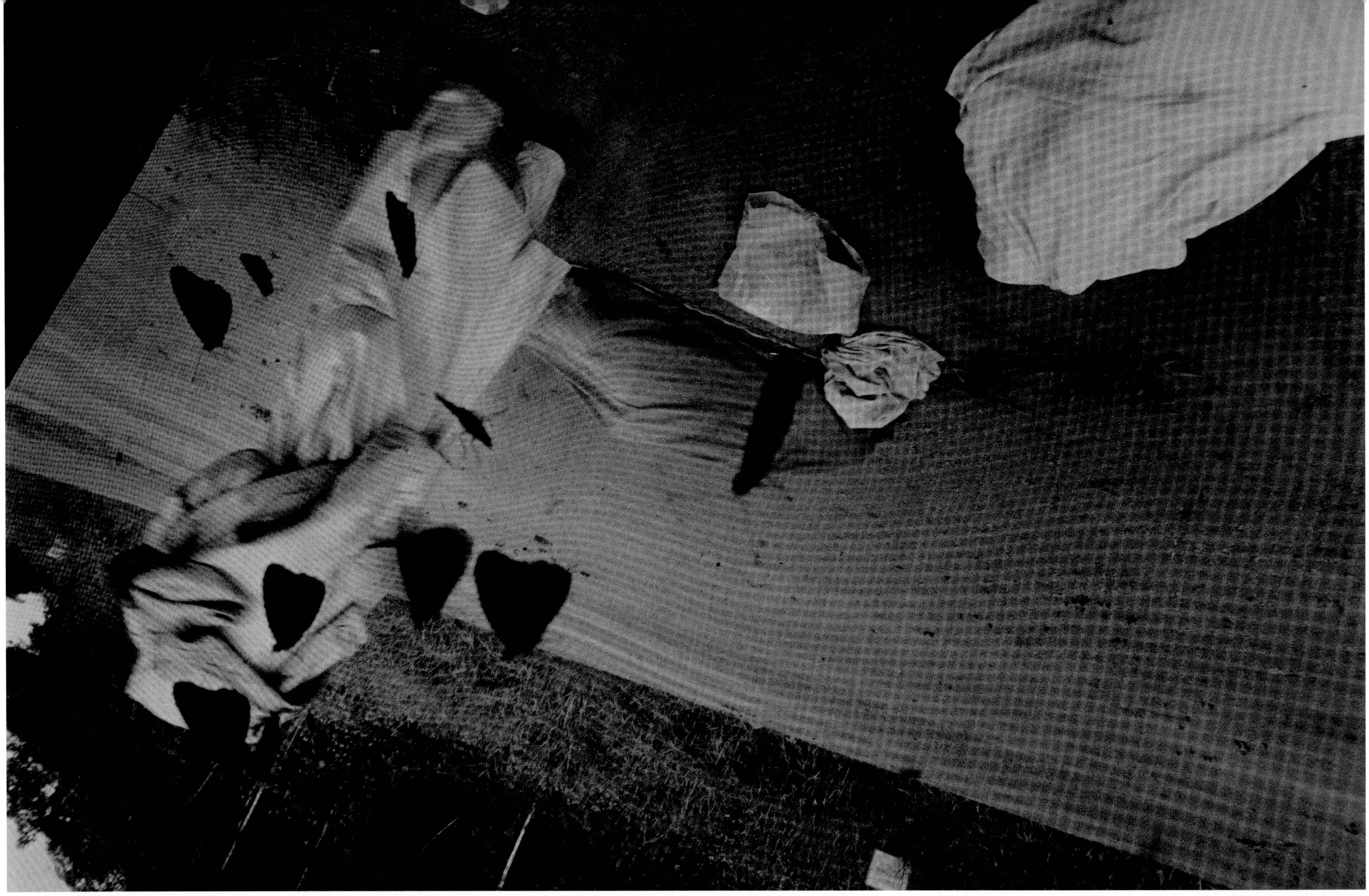

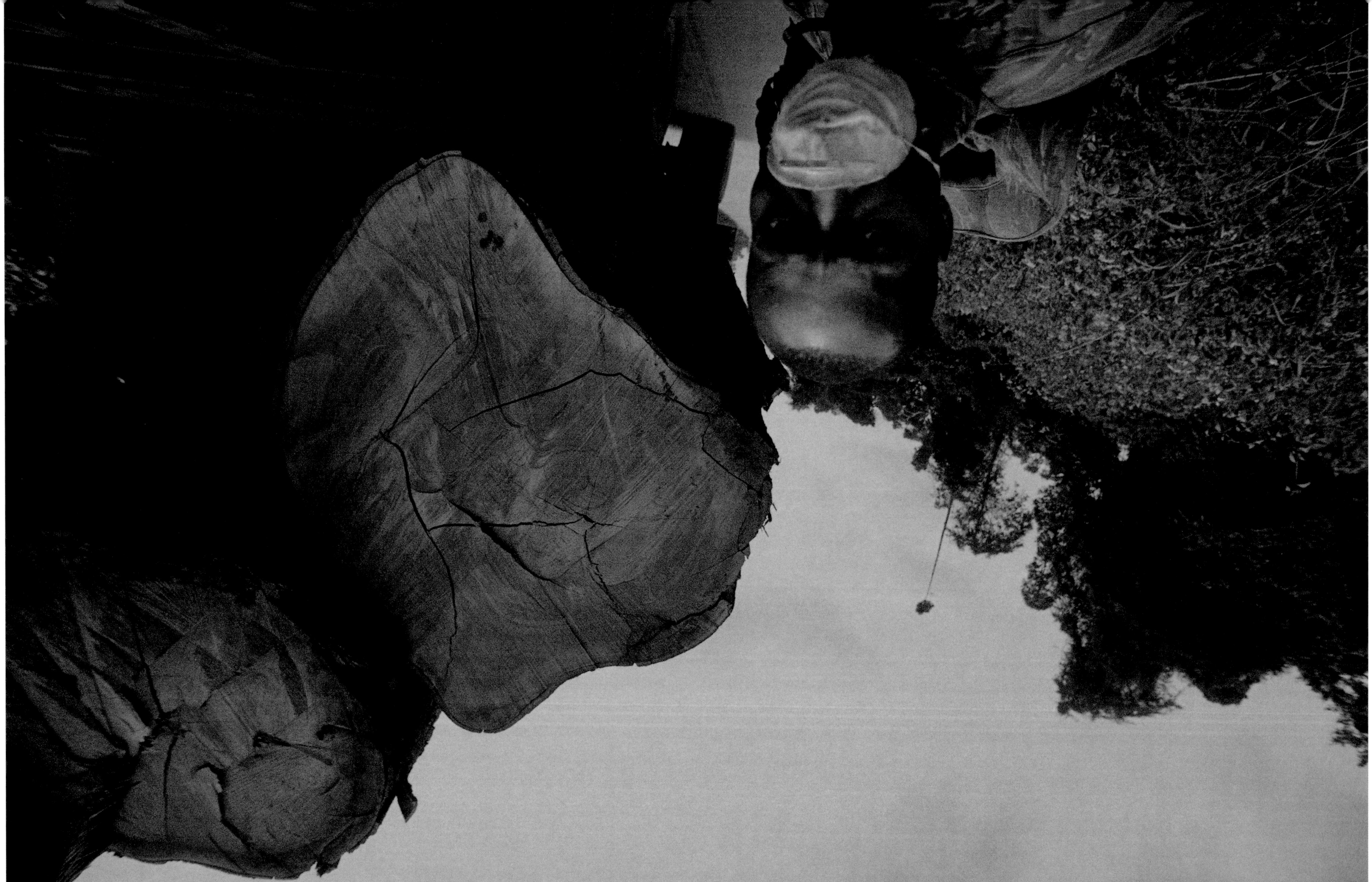

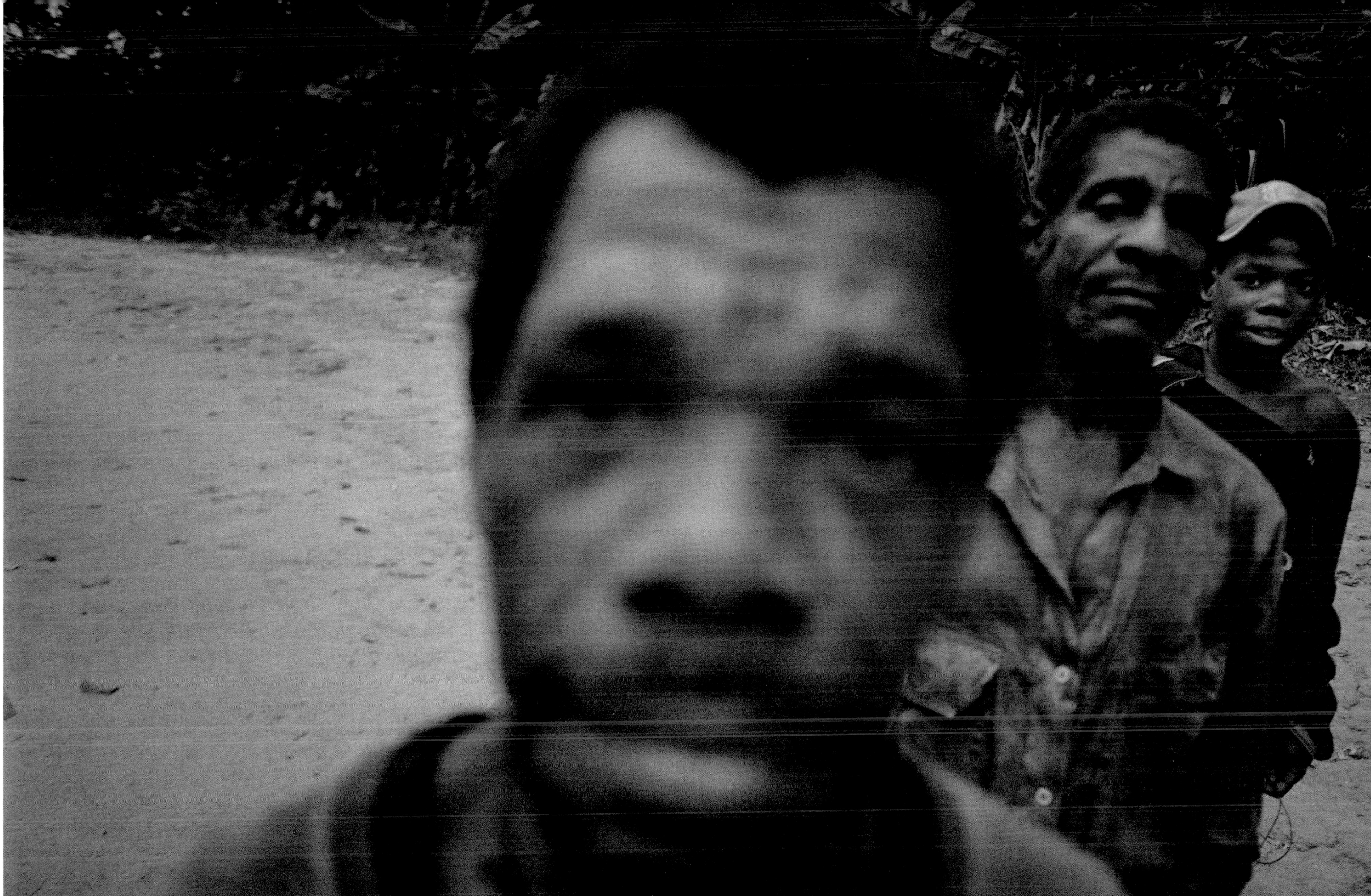

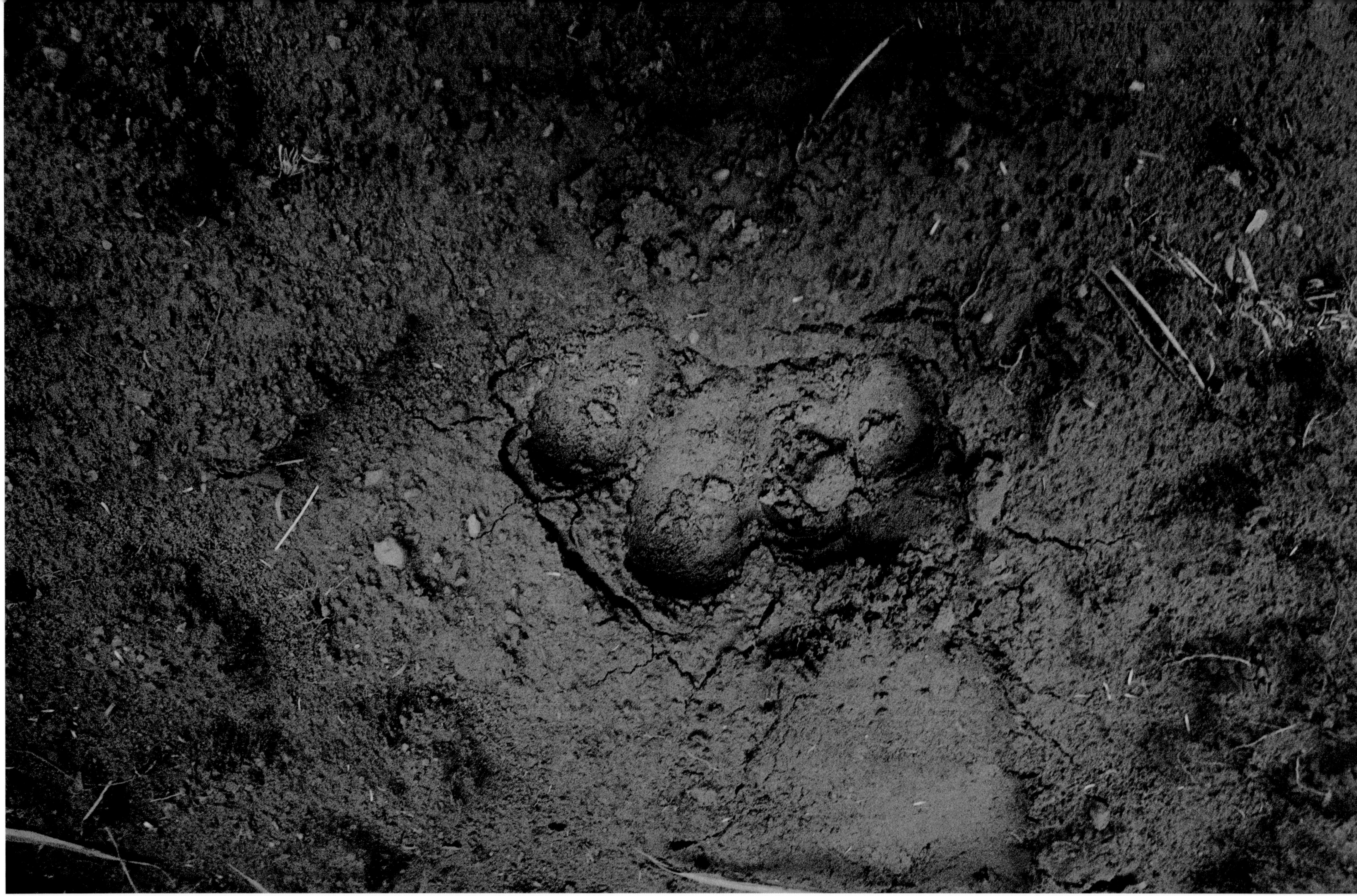

Today Morrisonville's cemetery lies within Dow Chemical's chain link fences.
Relatives of the dead require special permission to visit what they had free access to before the company took charge of the land ...
special permission from the chemical plant that denies any responsibility for the ailments.
of those citizens who lived there at one time still suffer with. It took the citizens of the town 30 years to make the connection.

Thirty years.
for this.
to all come out in.
tumors,
cancers ...
unnatural death.

Thirty years is a span of time that before had no name.
Dow changed that.

TODAY IT'S KNOWN AS [THE LATENCY PERIOD] :
THE TIME IN WHICH IT TAKES A PERSON TO SHOW THE PHYSICAL EFFECTS OF LONG TERM EXPOSURE TO *environmental pollution* .

No one lives in Morrisonville anymore, although, the number of residents a decade ago touched 300.
Three-hundred, a church and one cemetery.
That doesn't seem much to anyone looking in, but then again,
for a town founded by slaves in the late 1700s that counts for something,or at least it did.
It's different today.

MORRISONVILLE.
nestles beside the banks of the Mississippi River.
where it takes one of its many meandering,
long, looping bends that never quite make the figure eight.
Morrisonville is connected to the slightly larger town of Plaquemine,
that's French for persimmon,
a delicate fruit that's very sweet when ripe.
The succeeding generations of those slaves lived here self-sufficiently,
growing their own food, fishing the levees,
establishing themselves and their families' futures permanently.
Life was sweet in those days, sweet like a plaquemine.

In 1958, **DOW CHEMICAL** came to Morrisonville and opened its Plaquemine plant.
With it, came the spaghetti systems of spigots and pipes needed to refine and distill petroleum.
The company erected 300 foot stacks that sent 100 foot flames "flaring" at all hours, creating daytime in the dead of the night.
Drawn by the promise of cheap overhead,
access to raw materials and the Mississippi for cooling machines and providing a nearby dump,
the company began producing its plastics and other oil-based products.
The state of Louisiana said Dow Chemical would be good for local business.
Morrisonville never saw any of that.

Soon after the plant opened [DAISIES] appeared.
Every home in Morrisonville had them.
These weren't the kind of flowers residents tended like they had their crops.
These weren't flowers at all. Daisy Boxes were loud speakers,
Dow's equivalent of a canary in a coal mine ...

They were important the company said.

The pellets,
liquids and.
powders.
that Dow makes produce other products;
bi-products like,
vinyl acetate,
benzene,
styrene and
chloroprene along with.
a whole host of other sulphuric chemicals that are
discharged into the air and
water on a daily basis.

Moreover, the bi-products are injected underground inside the plant's perimeters.
For the townspeople
whose gardens backed up to the plant,
the millions of pounds of toxic releases per year
set in motion something invisible
that would eventually come out:
[THE LATENCY PERIOD] was underway.

By the late eighties the company began a program of relocation.

IN ESSENCE , THE PROGRAM WAS THE REMOVAL OF TOWNSPEOPLE FROM THEIR HOMES ... NOT ORDERED ... AIDED .

louisiana _ when the chemical corridor became cancer alley ...

_ *Section*

VI

WHEN THE CHEMICAL CORRIDOR BECAME CANCER ALLEY ...

Dow offered money,
incentives and
newly built residences in towns further away
so that the inhabitants wouldn't have to
put up with the industry's bangs, hisses and clatter.
In some instances the deals were so sweet people came out with more than they'd started with.
The company was generous with their promises of health benefits and
college scholarships for
those willing to leave-generous indeed.
AS THE TOWN STARTED THINNING, THE CEMETERY STARTED FILLING.

Today, only a couple of homes and the church named after preacher Morrison himself remain.
The rest is gone.
Four years before the arrival of Dow, in 1954,eight family run grocery stores were open for business on Railroad Avenue, now there's one.
The Dow Store is open though, selling Dow products cheaply to Dow employees.
Dow Chemical is the new plantation in these parts.
It took over from the Diamond plantation that ended its work practices in a time when human liberty was much talked about and later fought for.
The Civil War heralded better times, justice and equality for all.
Dow is a product of this freedom, acting freely towards those who had no say but had freedom.

Most regret leaving Morrisonville.
The induced migration hasn't been what Dow's pamphleteers had promised.
A nice home away from home never replaces community and,
money won't buy back histories,
histories that exist now in memory alone.
Like their ancestors who founded Morrisonville the townspeople have started again, only this time separated from their kin.
For Dow their migration policy has cost them a fraction of what any negligence lawsuit may have been.
The relocation program has been a success.
With every house torn down one less threat fell with it, one less legal time bomb removed.

None can say for certain what has caused the unusually high incidence of illness that has stricken the former residents of Morrisonville.

THERE IS NO CONNECTION
BETWEEN THE VARIOUS CANCERS,
ORGAN FAILURES AND
RESPIRATORY PROBLEMS
THEY SUFFER
VERSUS THE TOXIC EMISSIONS FROM THE
CHEMICAL PLANT.

The fact that there is an entire generation who have grown up never breathing clean air or
drank uncontaminated water has no medical bearing say Dow's attorneys.

A chemical consultant working for a local citizen's action group predicts that with
reduced emissions and more responsible disposal methods
the families of those who
built something unremarkable will one day regain their health ... not today ... in a few generations.

_ *Section*

VI

WHEN THE CHEMICAL
CORRIDOR
BECAME CANCER ALLEY ...

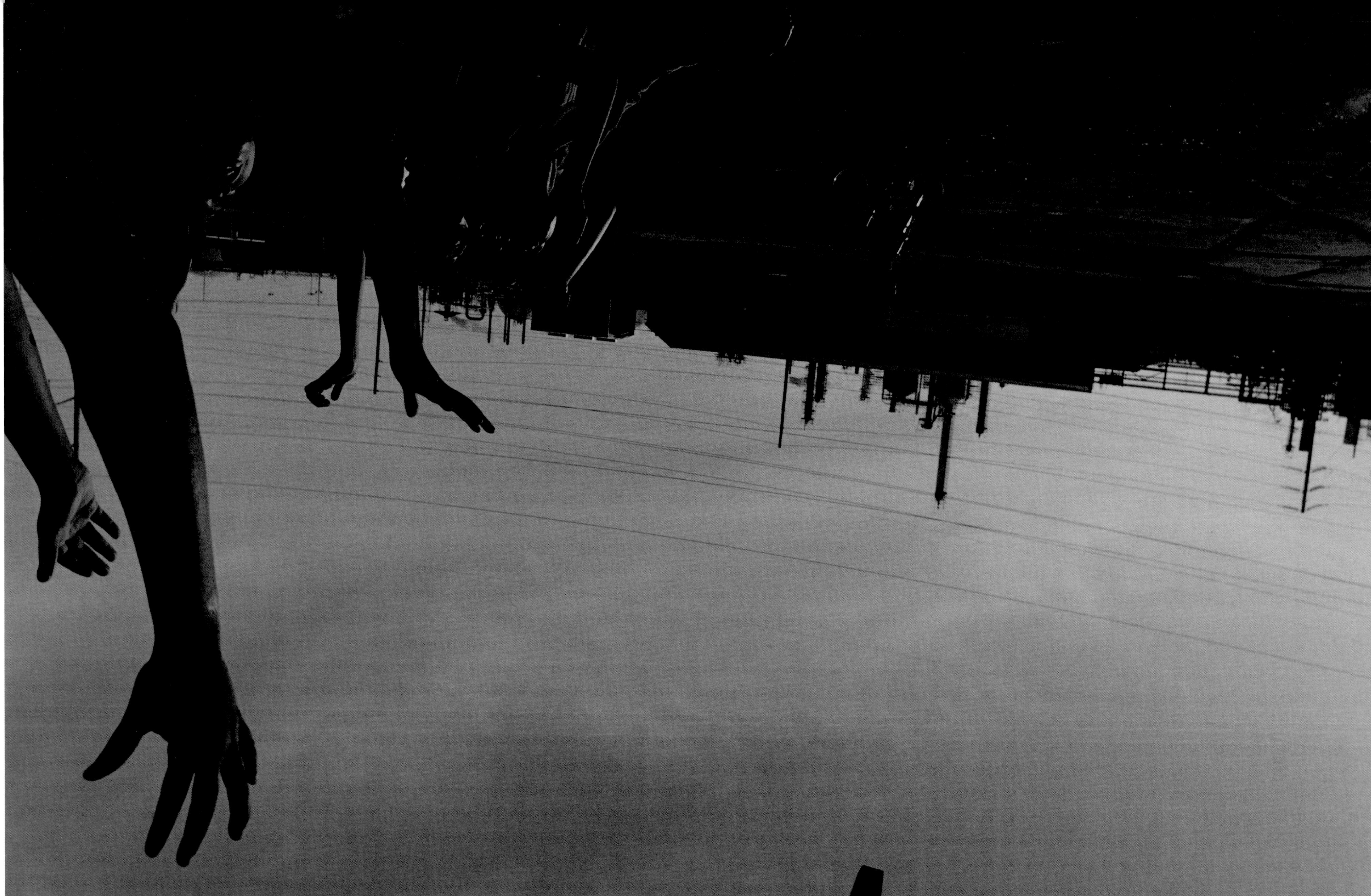

_ 091 BASKETBALL COURT NEXT TO CHEMICAL PLANT . BATON ROUGE

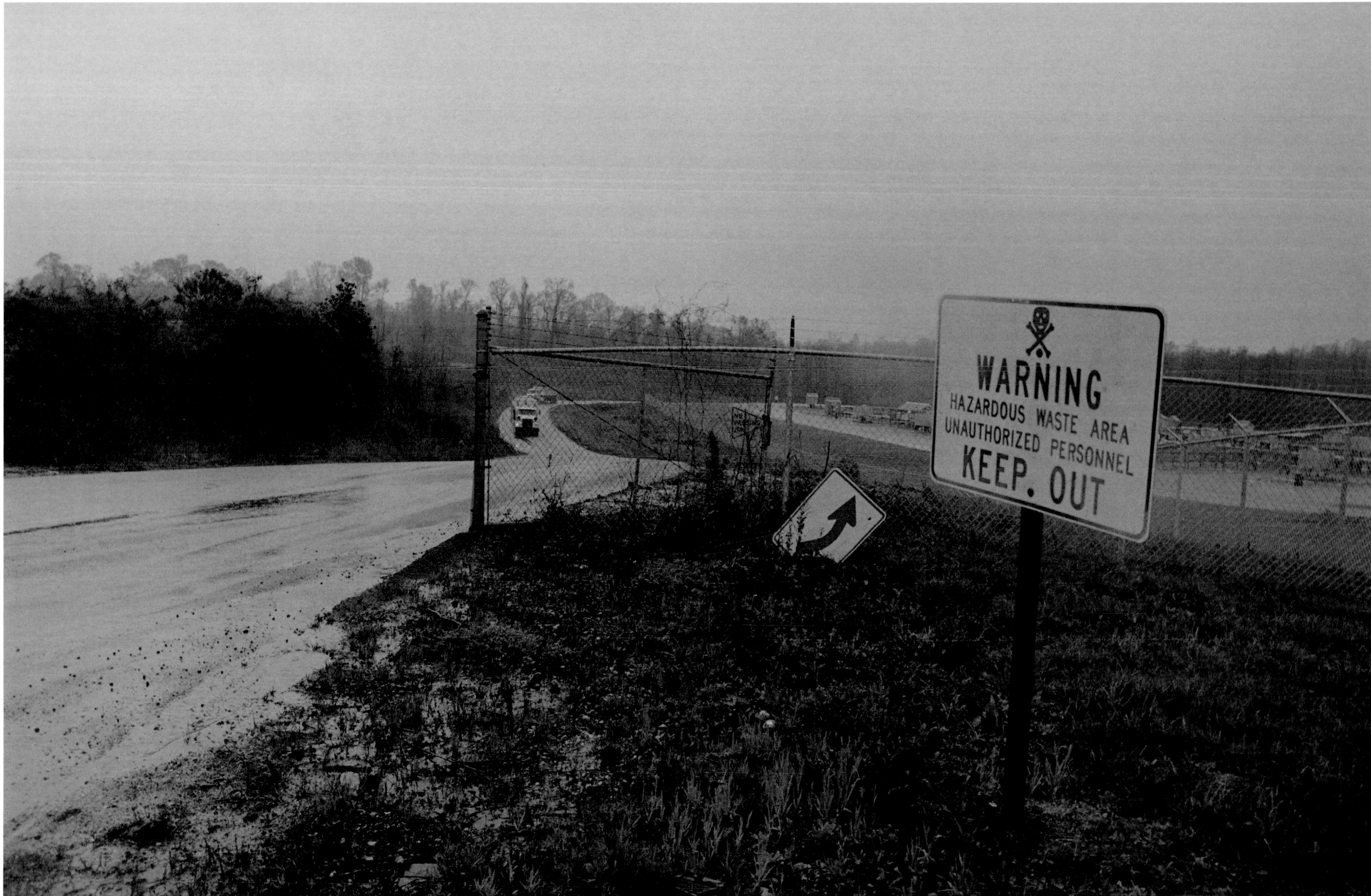
WARNING
HAZARDOUS WASTE AREA
UNAUTHORIZED PERSONNEL
KEEP. OUT

_ 093 TOXIC WASTE AREA
_ 094 A RAZED HOUSE IN MORRISONVILLE
_ 095 ROAD TO DOW CHEMICAL PLANT

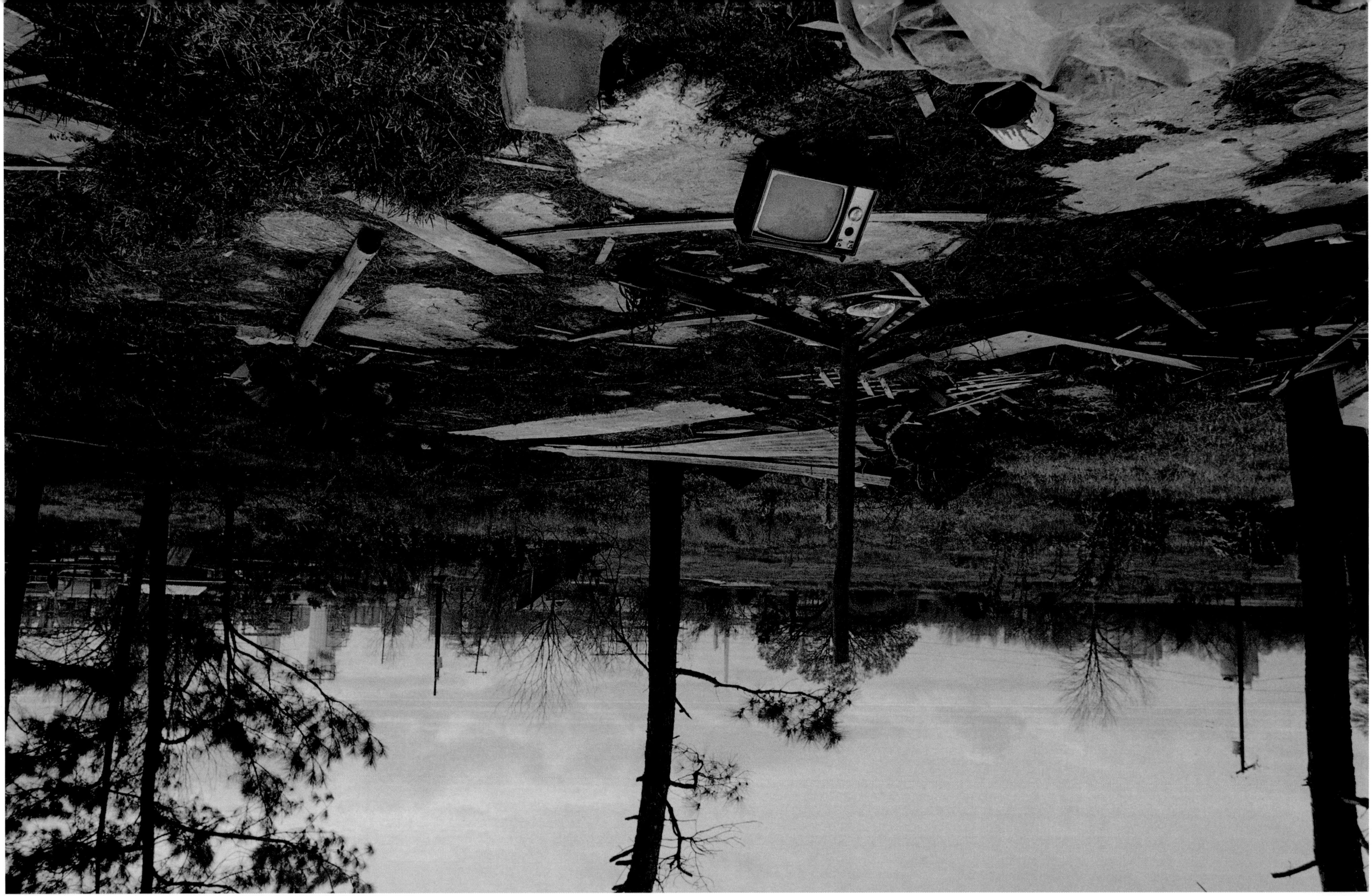

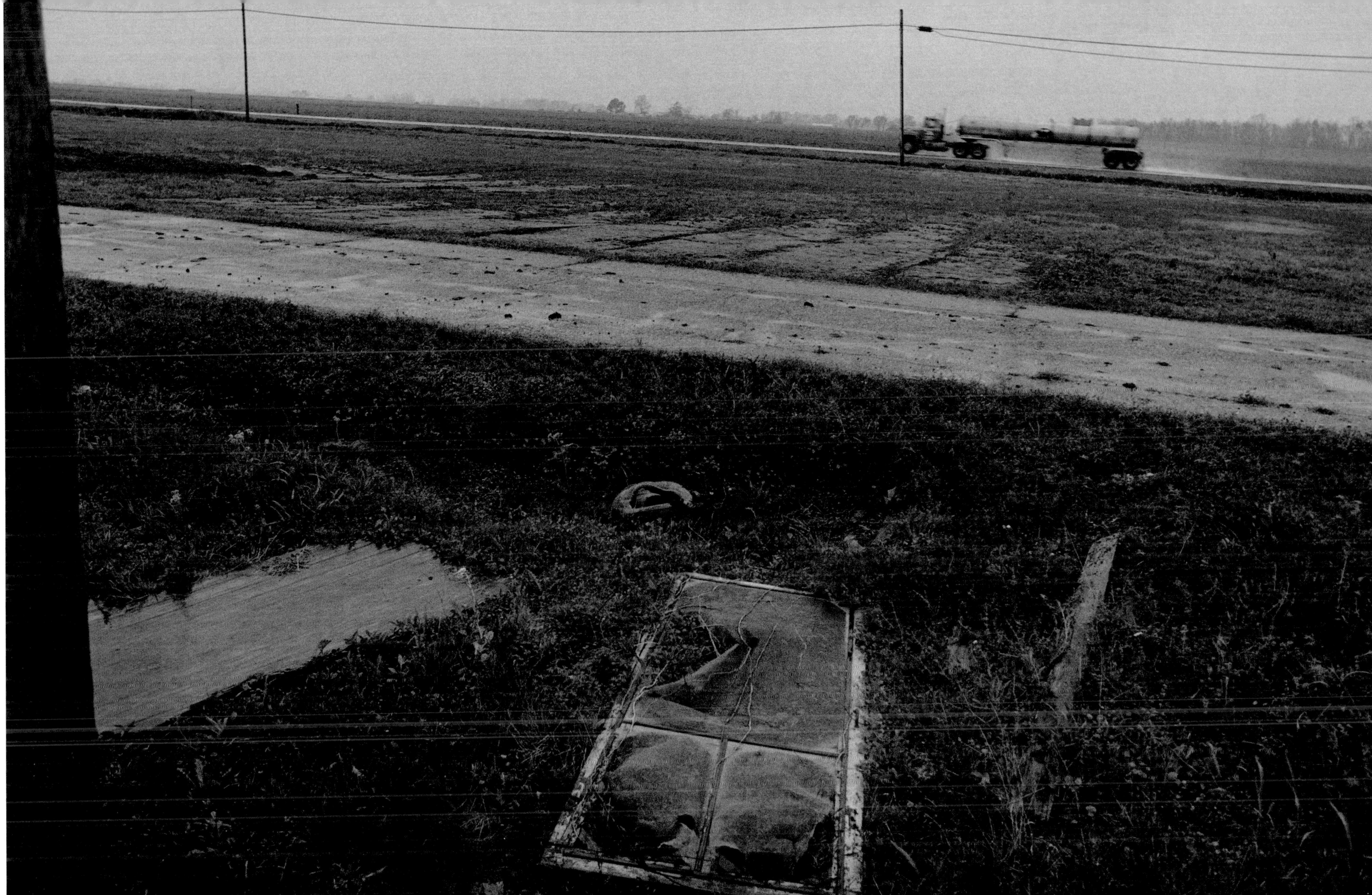

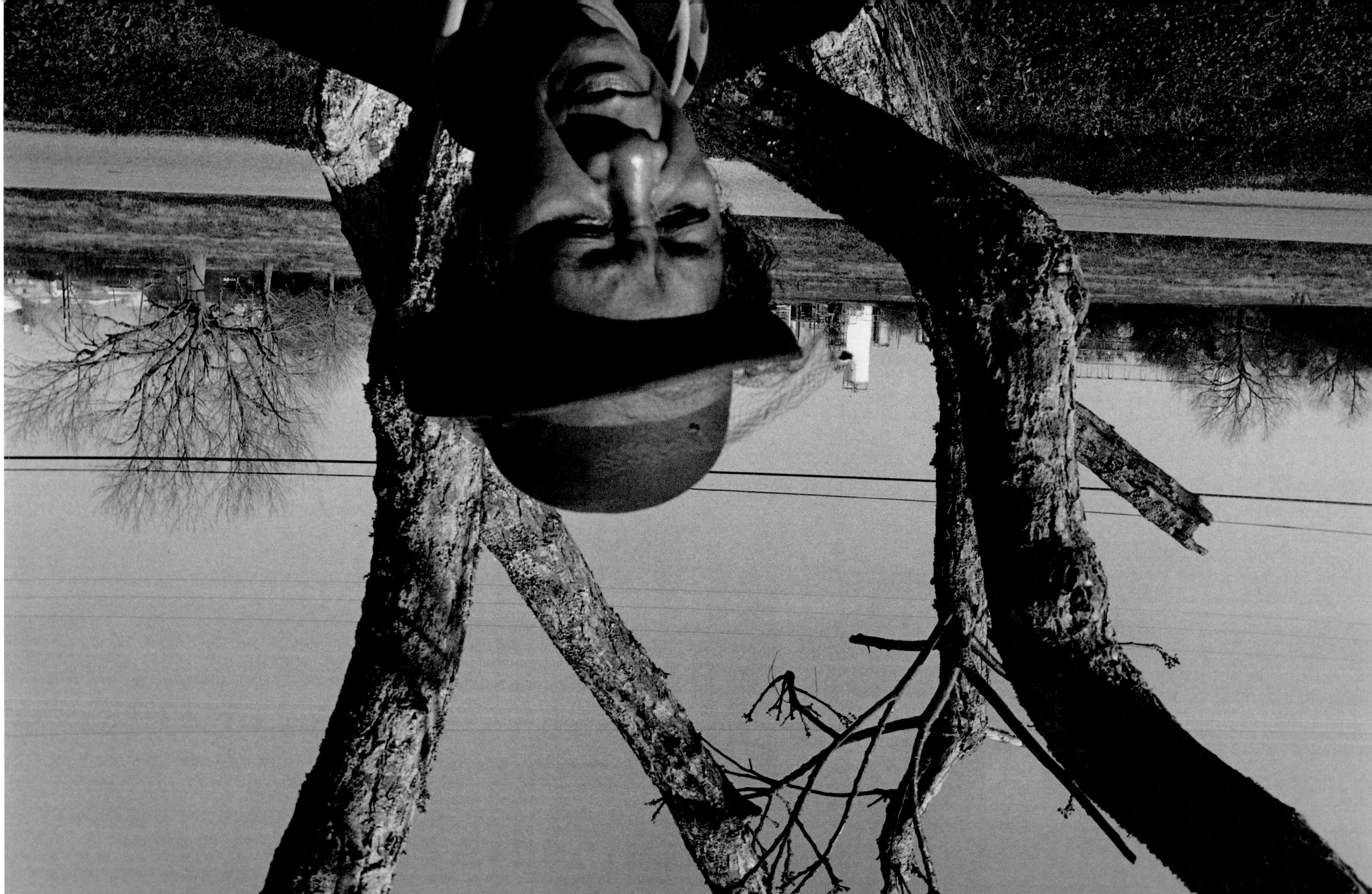

_ 097 A RESIDENT OF MORRISONVILLE STANDS WHERE HER HOME USED TO STAND

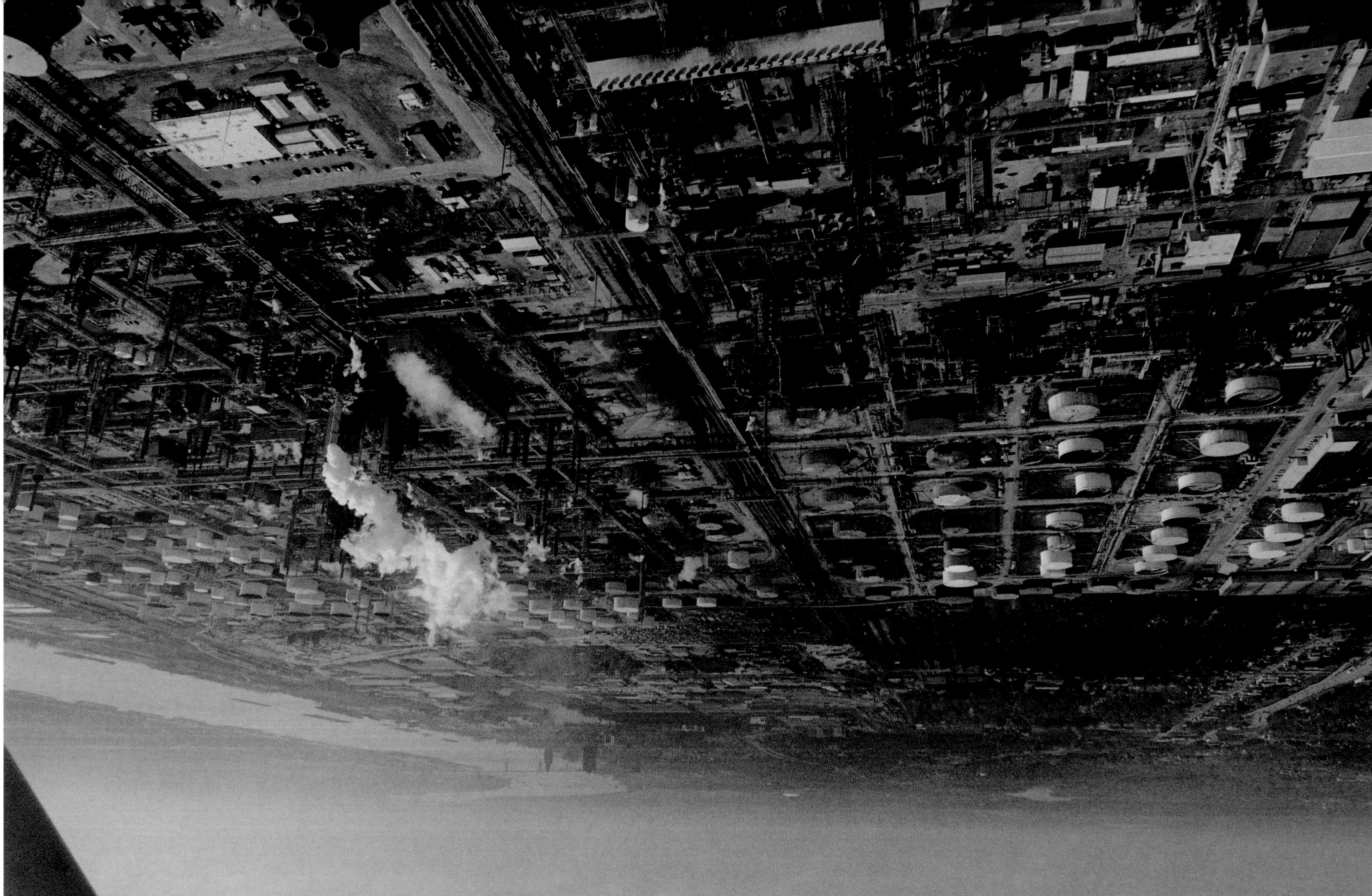

FROM RHODESIA,ZIMBABWE WAS BORN.

And with it the white African was brought a step closer to his end.
For those whites born under colonial rule to a queen with an empire, their country has long ceased to be.
For the subsequent generation of white Zimbabweans, their motherland has made them refugees.

These stateless,
landless
leftovers
of an expired venture
commenced by their forebears centuries ago
represent another page
in the list of human casualties.

FOR THE WHITE ZIMBABWEAN THE SLOW, INEVITABLE RESTORATION OF THE LAND TO THOSE WHO WERE DIVESTED OF IT IS BETRAYAL.

HE CONSIDERS HIMSELF AN AFRICAN, A NATIVE SON.

TO HIS BLACK COUNTRYMEN HE EMBODIES THAT WHICH SHOULD BE STRIPPED AND SENT AWAY.

The present has made a mockery of the past and
the white Zimbabwean a castaway in the spirit of another age.

zimbabwe _ not even their god was strong enough

_ *Section*

VII

NOT EVEN THEIR GOD WAS STRONG ENOUGH

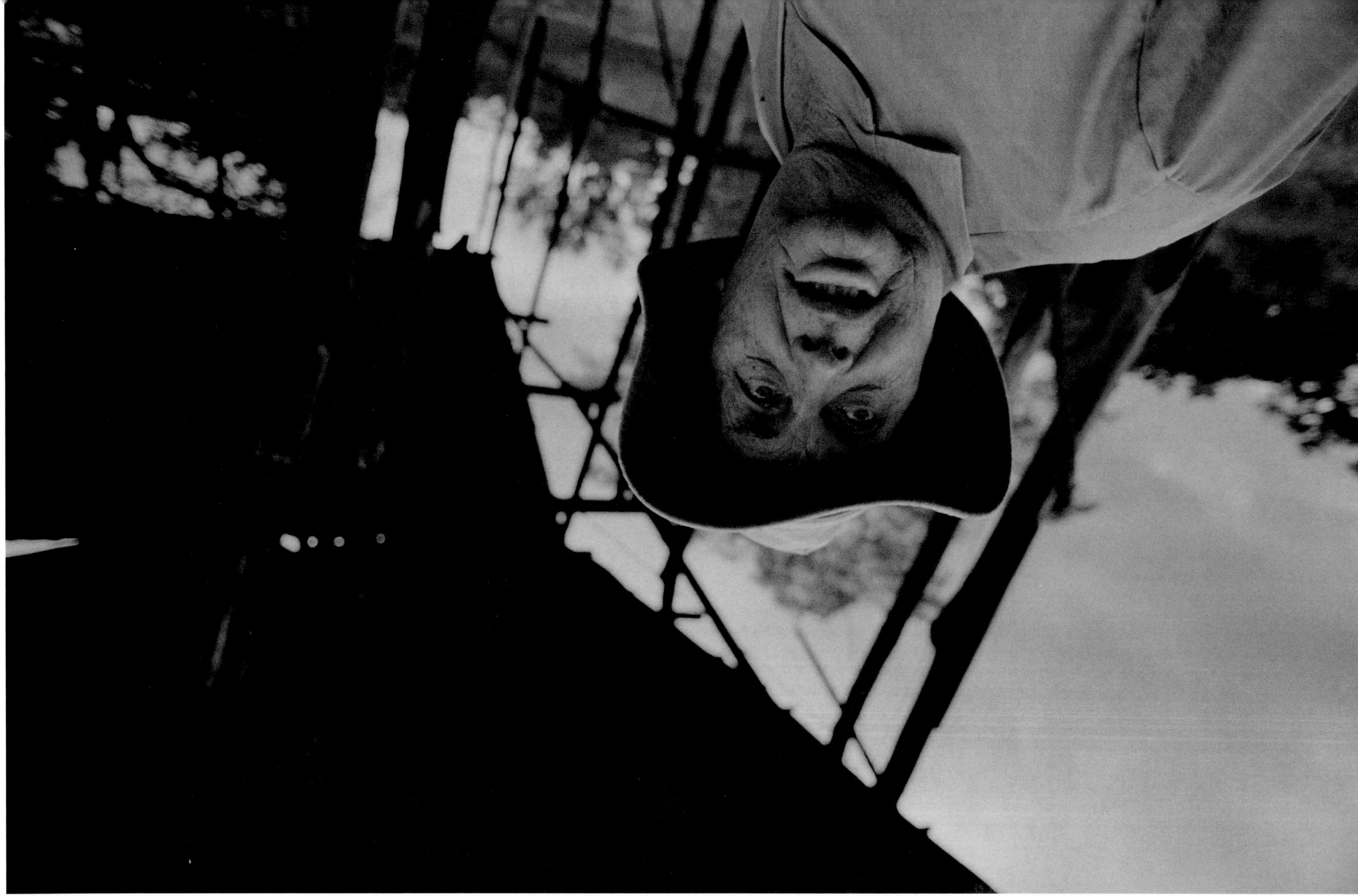

_ 103 FARMER IN FRONT OF HIS BARN BURNED DOWN BY WAR VETERANS
_ 104 WHITE FARMER'S HOUSE RANSACKED BY WAR VETERANS
_ 105 WHITE FARMER PICKS THROUGH THE REMAINS OF HIS RANSACKED HOME
_ 106 WAR VETERANS
_ 107 A WHITE FARMER'S DECLARATION OF COMPLIANCE WITH MUGABE'S NEW LAND POLICY
_ 108 POLITICAL GRAFFITI ON FARM HOUSE
_ 109 TOBACCO WORKERS ON WHITE FARM

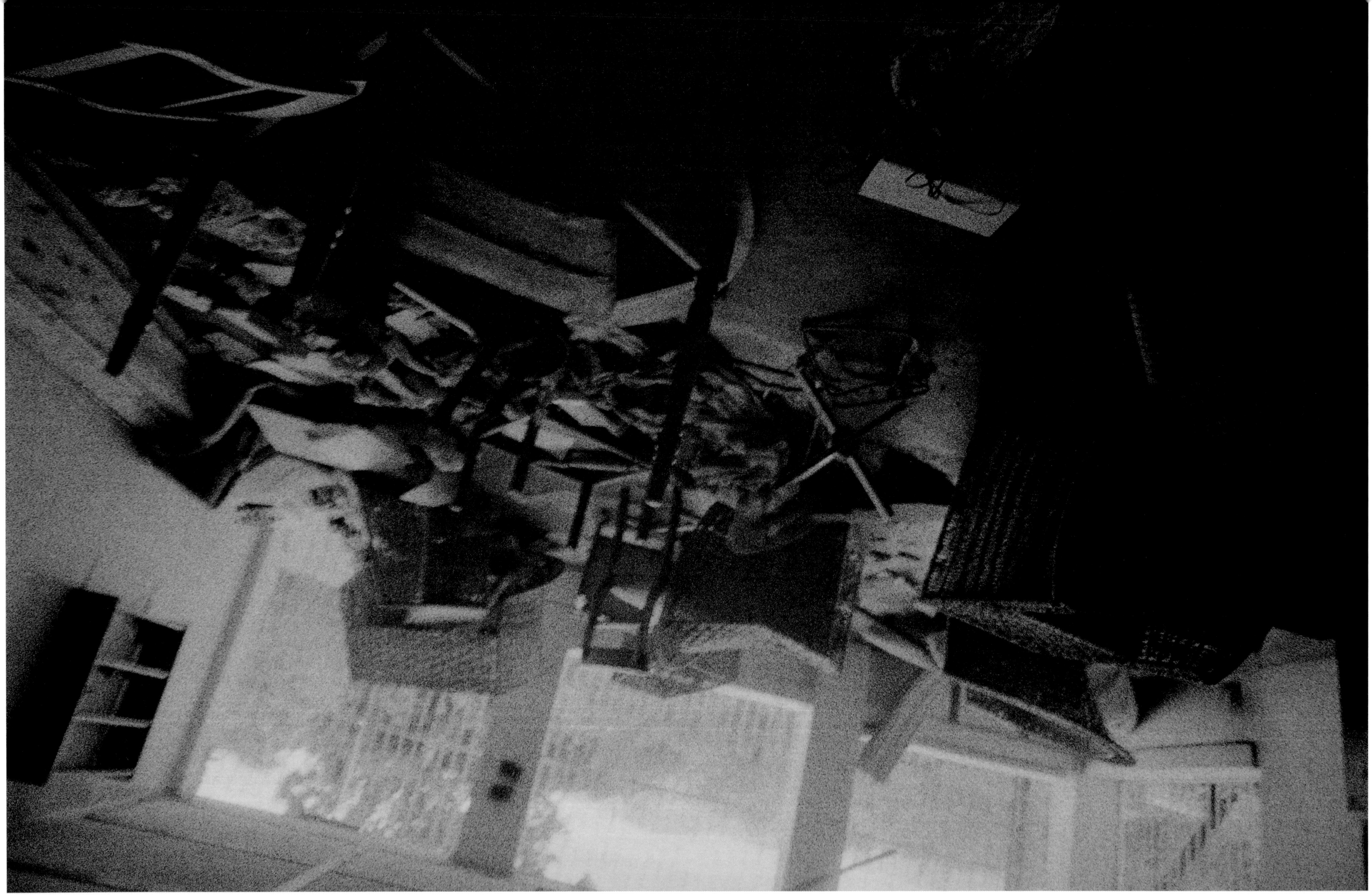

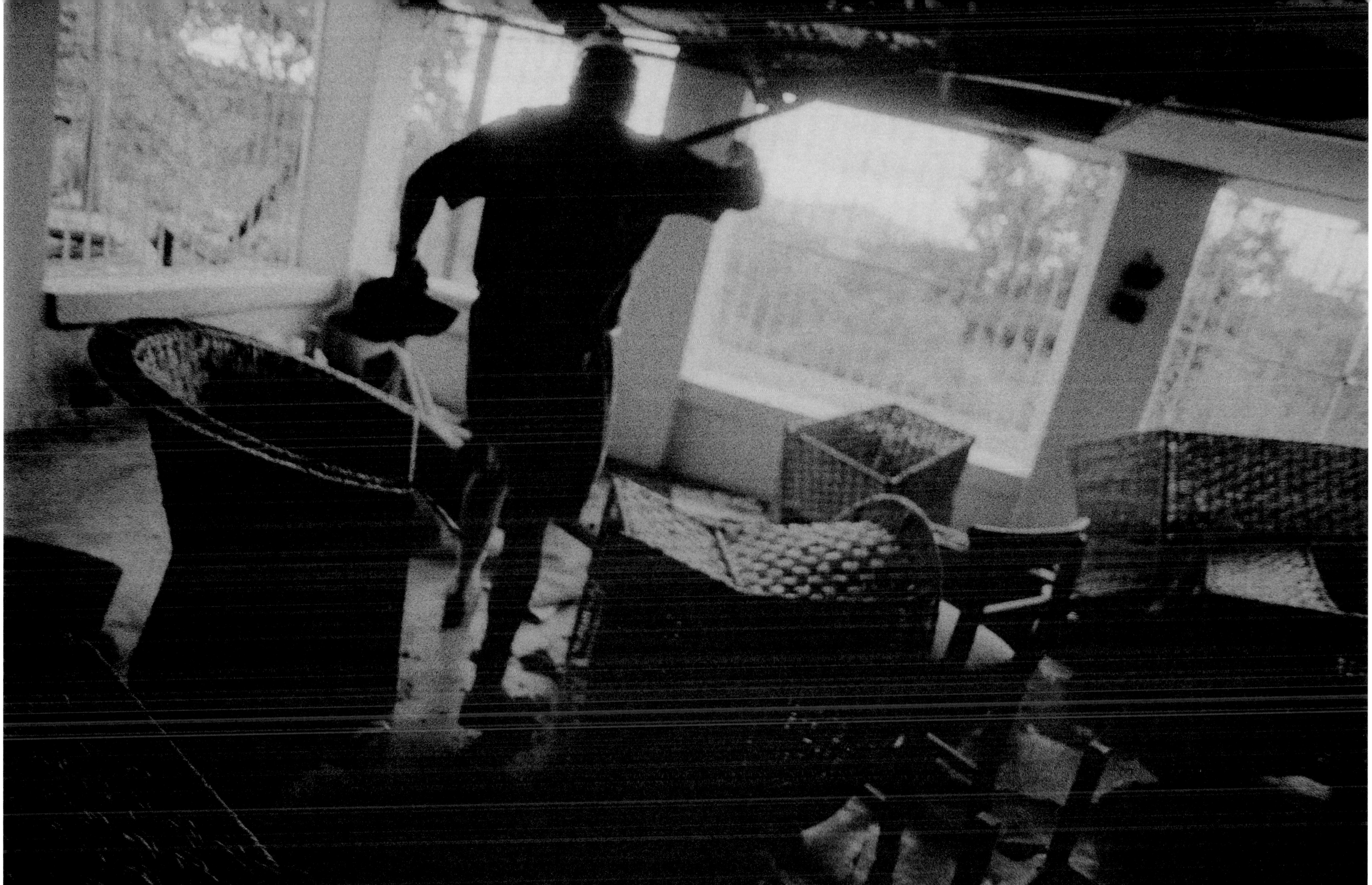

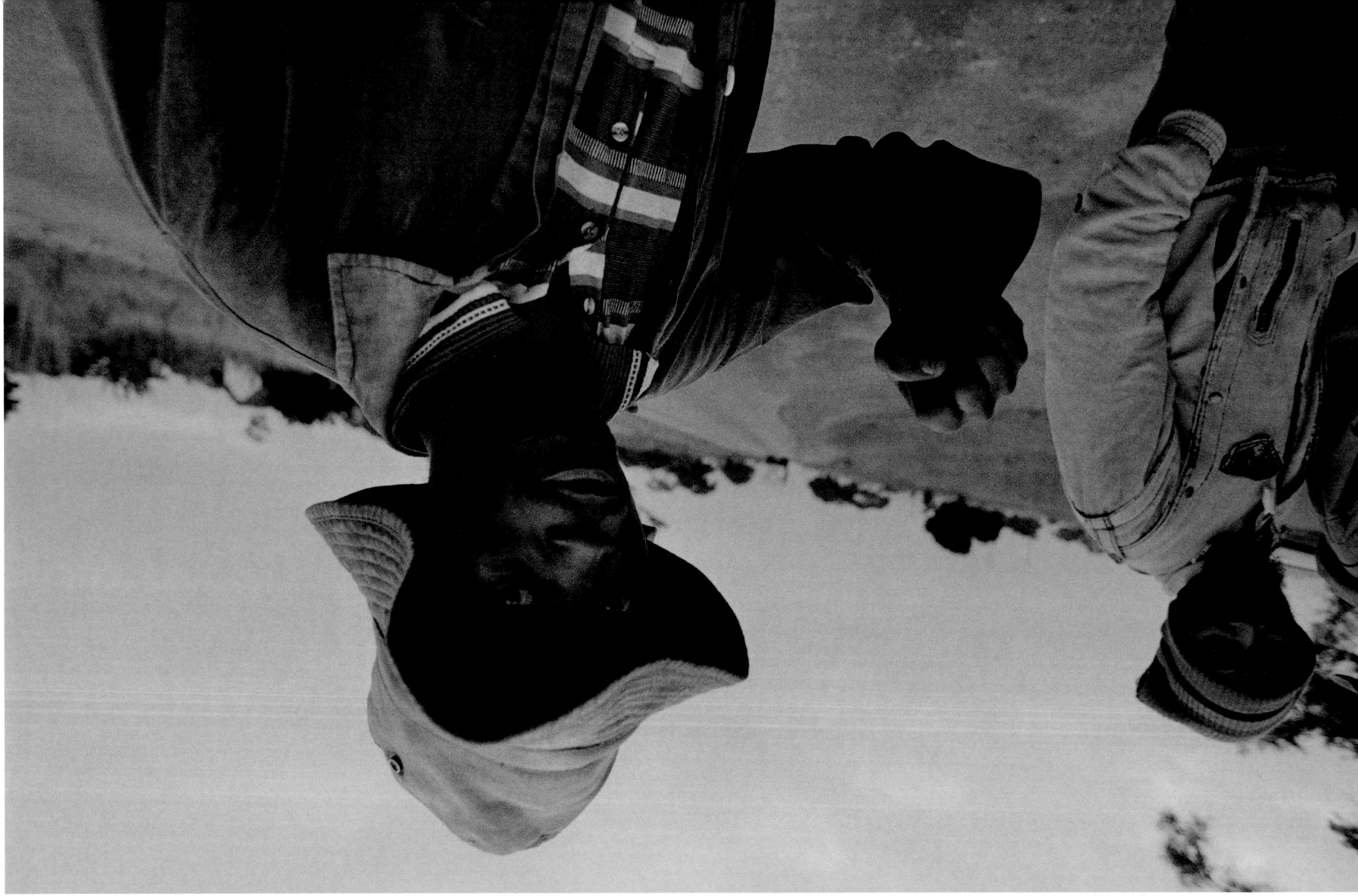

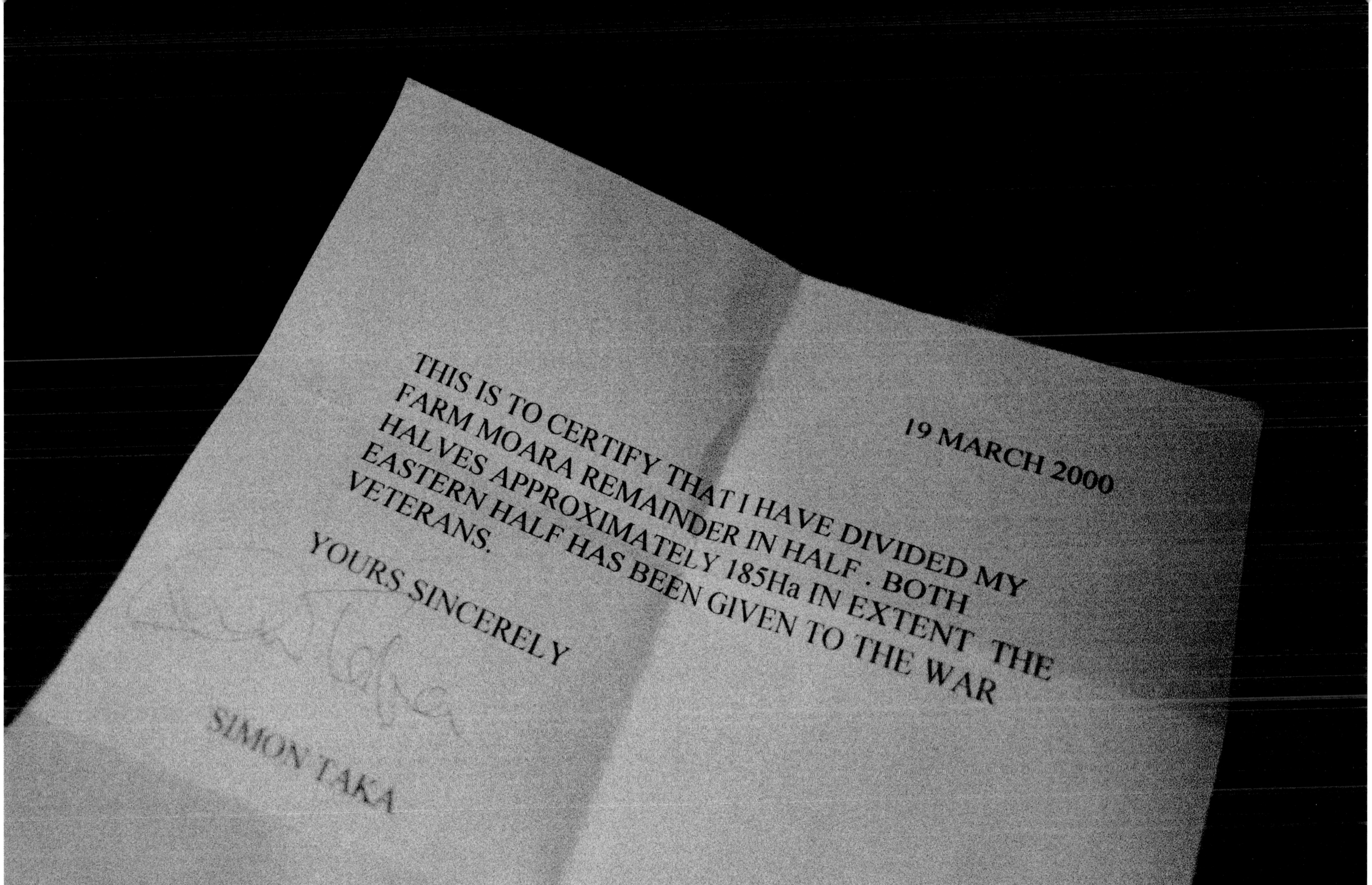

19 MARCH 2000

THIS IS TO CERTIFY THAT I HAVE DIVIDED MY FARM MOARA REMAINDER IN HALF . BOTH HALVES APPROXIMATELY 185Ha IN EXTENT THE EASTERN HALF HAS BEEN GIVEN TO THE WAR VETERANS.

YOURS SINCERELY

SIMON TAKA

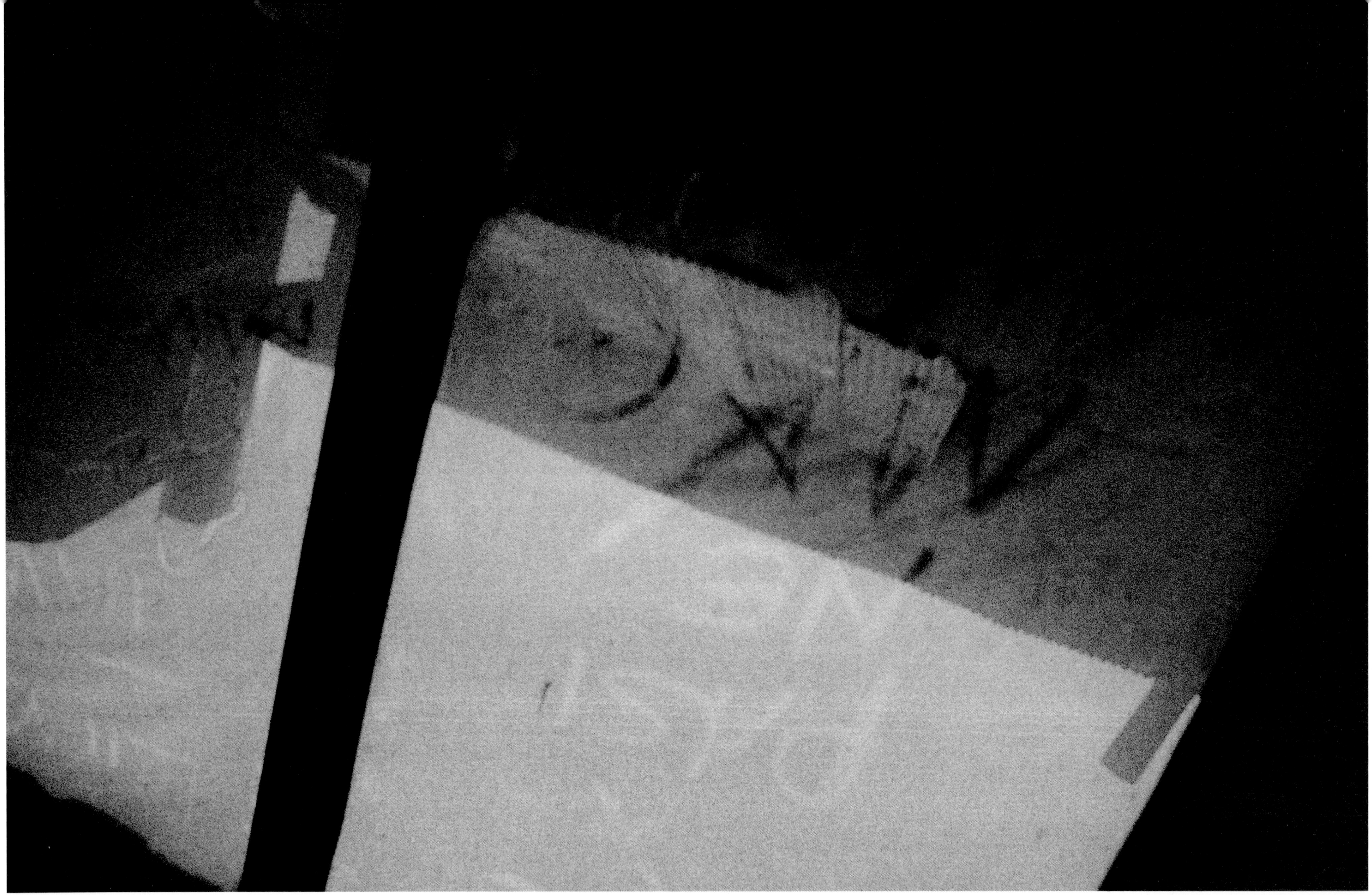

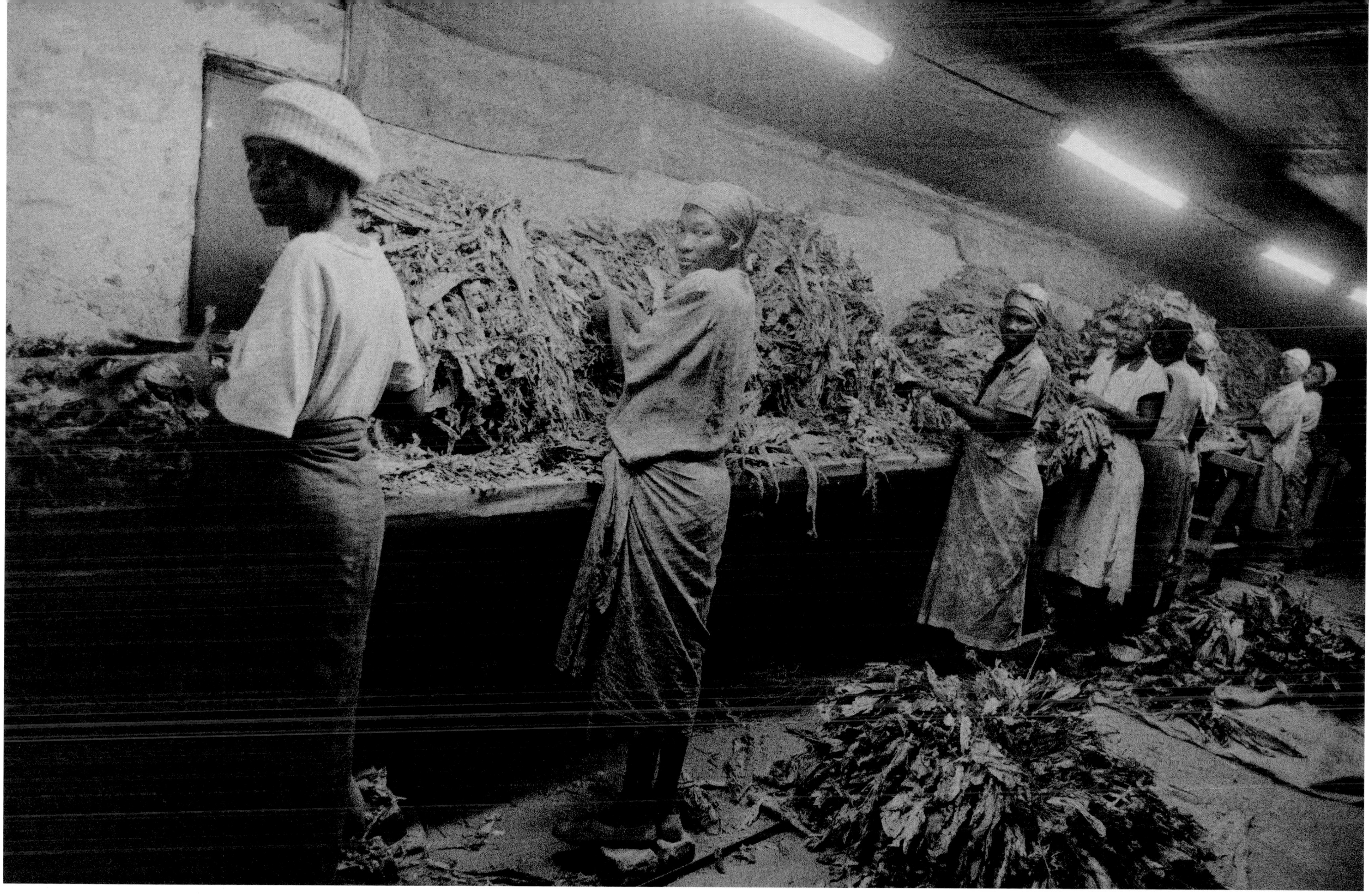

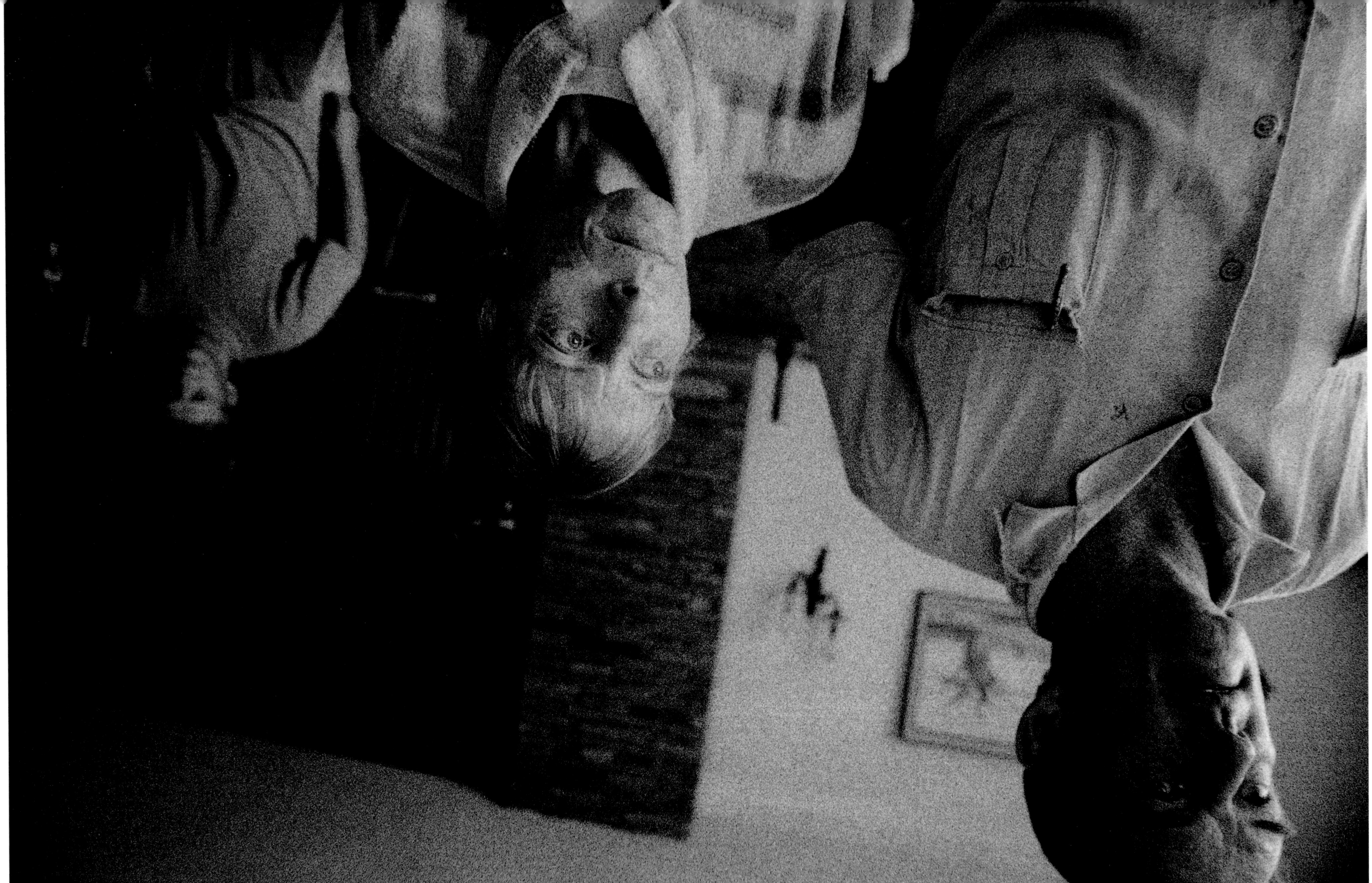

_ 111 ANXIOUS WHITE FARMERS
_ 112 DESTROYED FARM
_ 113 WHITE FARMER' DESTROYED TOBACCO SHED

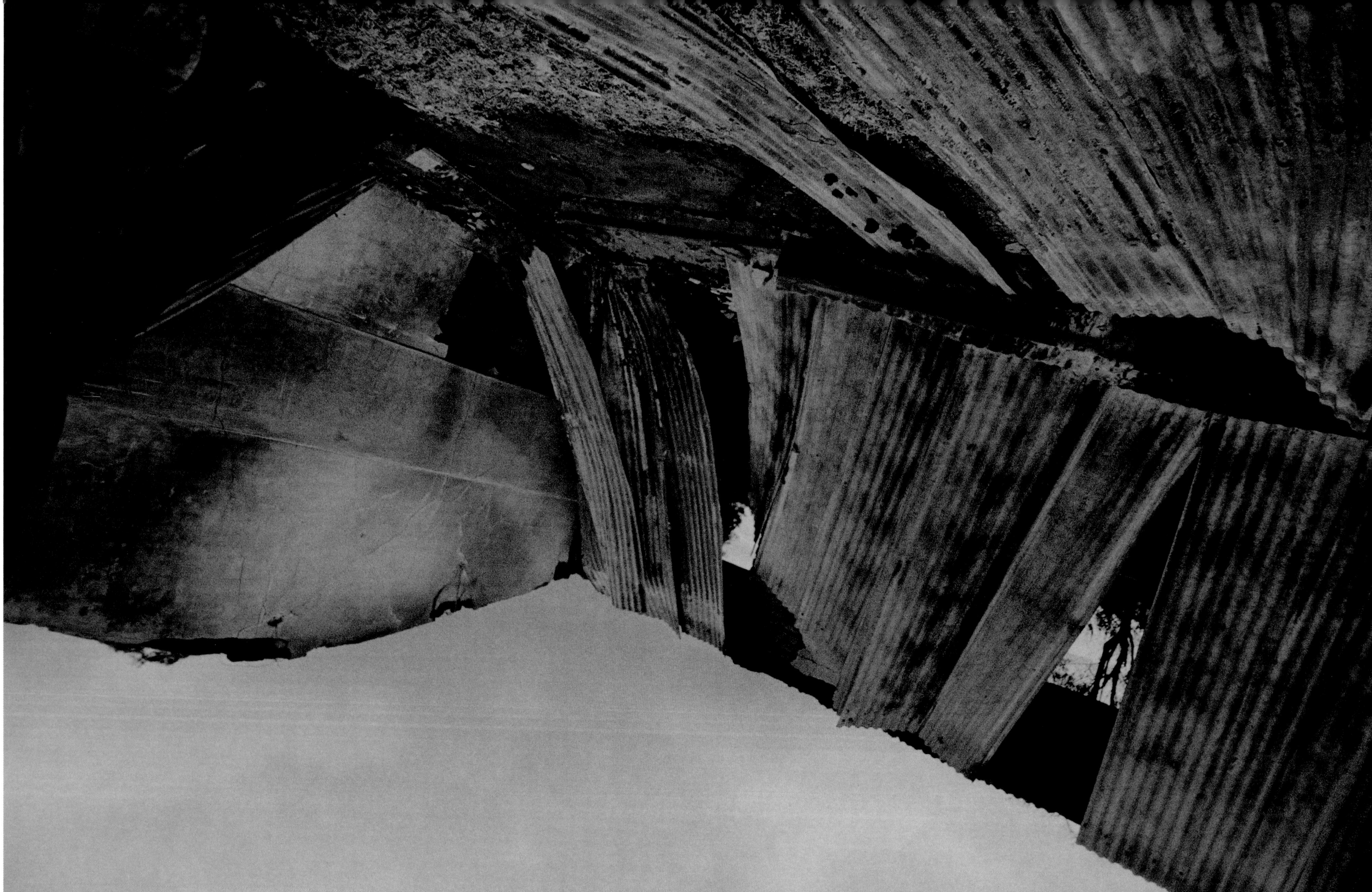

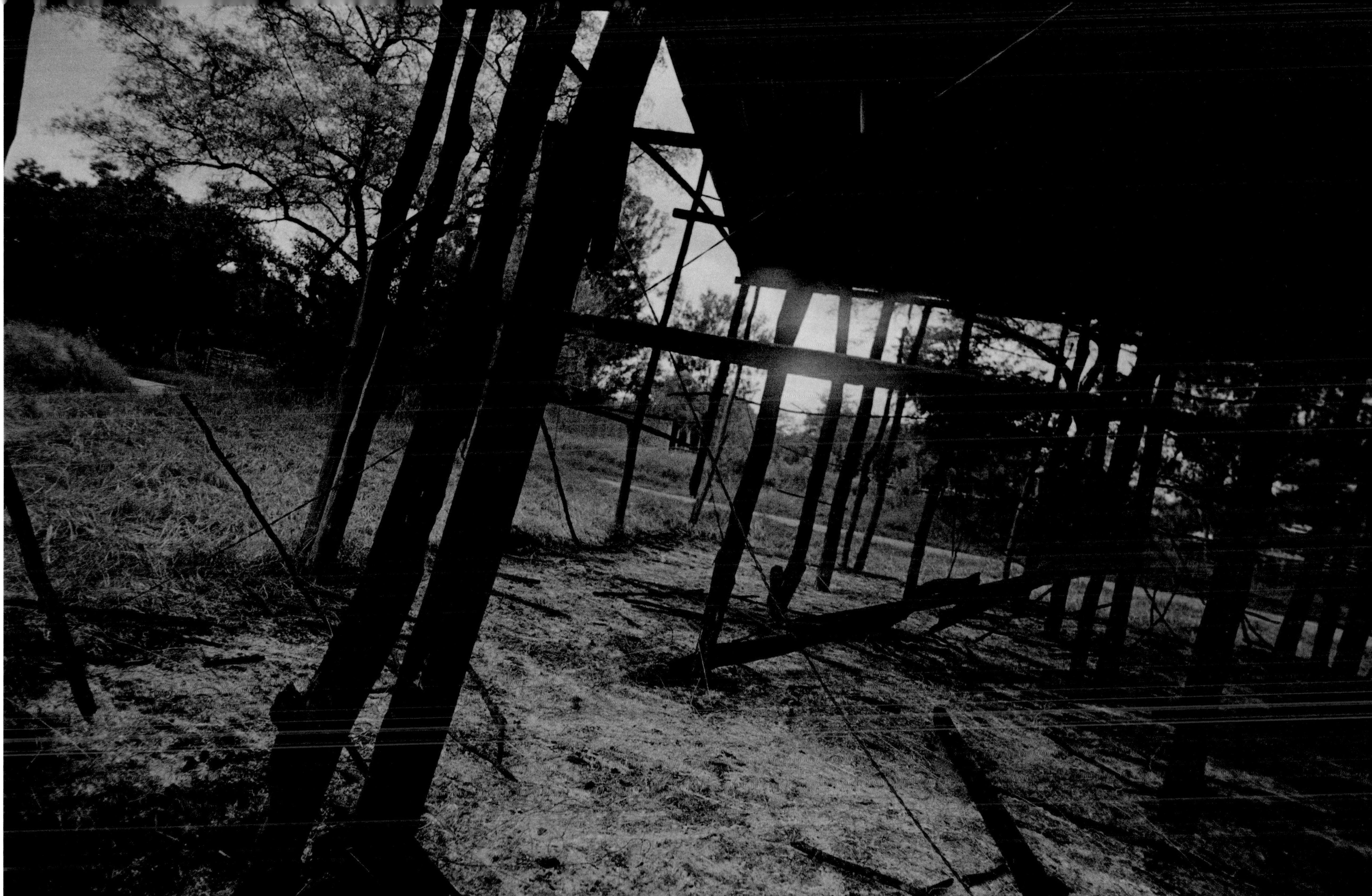

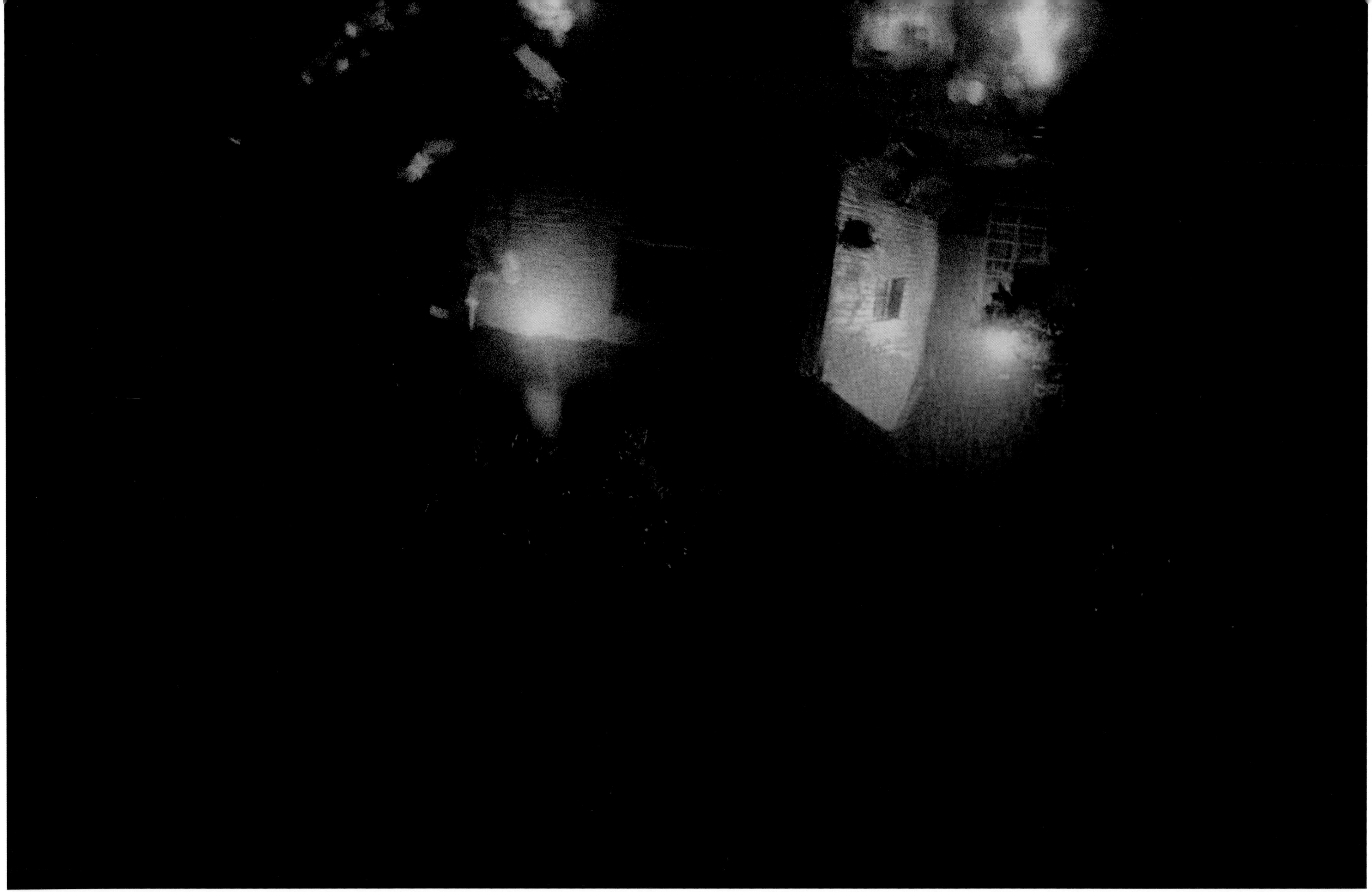

* TEXACO HISTORY TIMELINE:

1903. THE TEXAS COMPANY'S THIRD OIL WELL DRILLED AT SOUR LAKE, TEXAS, COMES IN AS A GUSHER, SAVING THE COMPANY FROM BANKRUPTCY.

Lago Agrio, meaning Sour Lake, was the place where Texaco Oil first struck it rich.

Lago Agrio was where the company first exploited the Ecuador's crude reserves and defiled its jungle.

* CHEVRONTEXACO EMPLOYEES WORK HARD TO ENSURE THAT OUR OPERATIONS AROUND THE WORLD.

ARE MANAGED IN A SAFE AND ENVIRONMENTALLY SOUND MANNER.

The company, it has been widely alleged, spilled 16.8 million gallons of oil.

(one and a half times that of the Exxon Valdez).

from ruptured pipelines, discharged 19 billion gallons of highly toxic water into waterways and the soil,

conducted massive deforestation, and left more than 600 toxic pits uncovered.

* TEXACO WEBPAGE-CURRENT ISSUES: DIVERSITY.

TEXACO, NOW A SUBSIDIARY OF CHEVRONTEXACO,HAS A STRONG COMMITMENT TO CREATING AND MAINTAINING A WORKPLACE THAT.

OPERATES ON THE FUNDAMENTAL PRINCIPLE OF RESPECT FOR THE INDIVIDUAL, AND WHERE EMPLOYEES ARE TREATED FAIRLY WITHOUT REGARD TO RACE,

RELIGION, COLOR, NATIONAL ORIGIN, AGE, SEX, SEXUAL ORIENTATION, DISABILITY, VETERAN STATUS OR POSITION WITHIN THE COMPANY.

The [indigenous] Cofan, who in 1971—

when ChevronTexaco first began operations on their land-numbered 15,000, have seen their population in the area fall to less than 300.

* "THERE HAS NEVER BEEN MADE ANY CREDIBLE, SCIENTIFIC EVIDENCE TO SUPPORT THE ALLEGATIONS MADE,"

SAID CHRIS GILDEZ,

A CHEVRONTEXACO SPOKESMAN ON THE.

TEXACO POLLUTION LAW SUIT.

"THE PLAINTIFFS HAVE OFFERED AS EVIDENCE ONLY PSEUDOSCIENCE."

... a 1987 study by the U.S. Environmental Protection Agency found that small wells that dumped no more than 100 barrels of waste water a.

day into. streams could slightly increase the risk of cancer among local residents.

This was the case in the U.S. In the Amazon,

Texaco was dumping up to 100,000 barrels of waste-water a day—1000 times more.

* TEXACO: ABOUT TEXACO: FAST FACTS-WHAT WE SUPPORT.

TEXACO HAS A LONG HISTORY OF SUPPORTING BOTH NATIONAL AND LOCAL PROGRAMS.

IN THE COMMUNITIES WHERE WE DO BUSINESS.

THESE INCLUDE A WIDE RANGE OF EDUCATIONAL,

ENVIRONMENTAL,

SOCIAL AND.

HEALTH CARE INITIATIVES.

"We can't keep living like worms in oil." *Secoya Indian leader Elias Piyahuaje.*

* TEXACO IN ECUADOR: LEGAL ARCHIVES.

ON MAY 30, 2001, JUDGE JED RAKOFF OF THE U.S. DISTRICT COURT IN NEW YORK DISMISSED THE LAWSUITS AGAINST TEXACO.

IN HIS RULING, JUDGE RAKOFF NOTED,

"BECAUSE TEXACO HAS CARRIED ITS BURDEN ON EVERY ELEMENT OF THE MOTION,

AND BECAUSE THE RECORD ESTABLISHES OVERWHELMINGLY THAT THESE CASES HAVE EVERYTHING TO DO WITH ECUADOR AND.

NOTHING TO DO WITH THE UNITED STATES."

This includes the decision to use unlined waste pits for production water.
Even expenses as low as $5,000 had to be approved at the U.S. headquarters.

*** TEXACO IN ECUADOR: 21 OCT 2003 PRESS RELEASE-SUMMARY OF CHEVRONTEXACO RESPONSE TO ECUADOR LAW SUIT.**
THERE IS NO BASIS FOR THE SUIT SINCE NO EVIDENCE HAS BEEN PRESENTED THAT CHEVRONTEXACO CAUSED ANY DAMAGE TO THE PLAINTIFFS.

Vincente Alban works a small piece of land right next to a toxic pit.
It was "cleaned" only last week by one of Texaco's subcontractors.
The cleanup consisted of filling the pit with earth that flooded Alban's land with crude oil and toxic water.
"All the water in this town is contaminated," says Alban.
Only rainwater is drinkable here. It's collected in old Texaco oil drums.

ecuador _ lords of the jungle

_ *Section*
VIII
LORDS OF THE JUNGLE

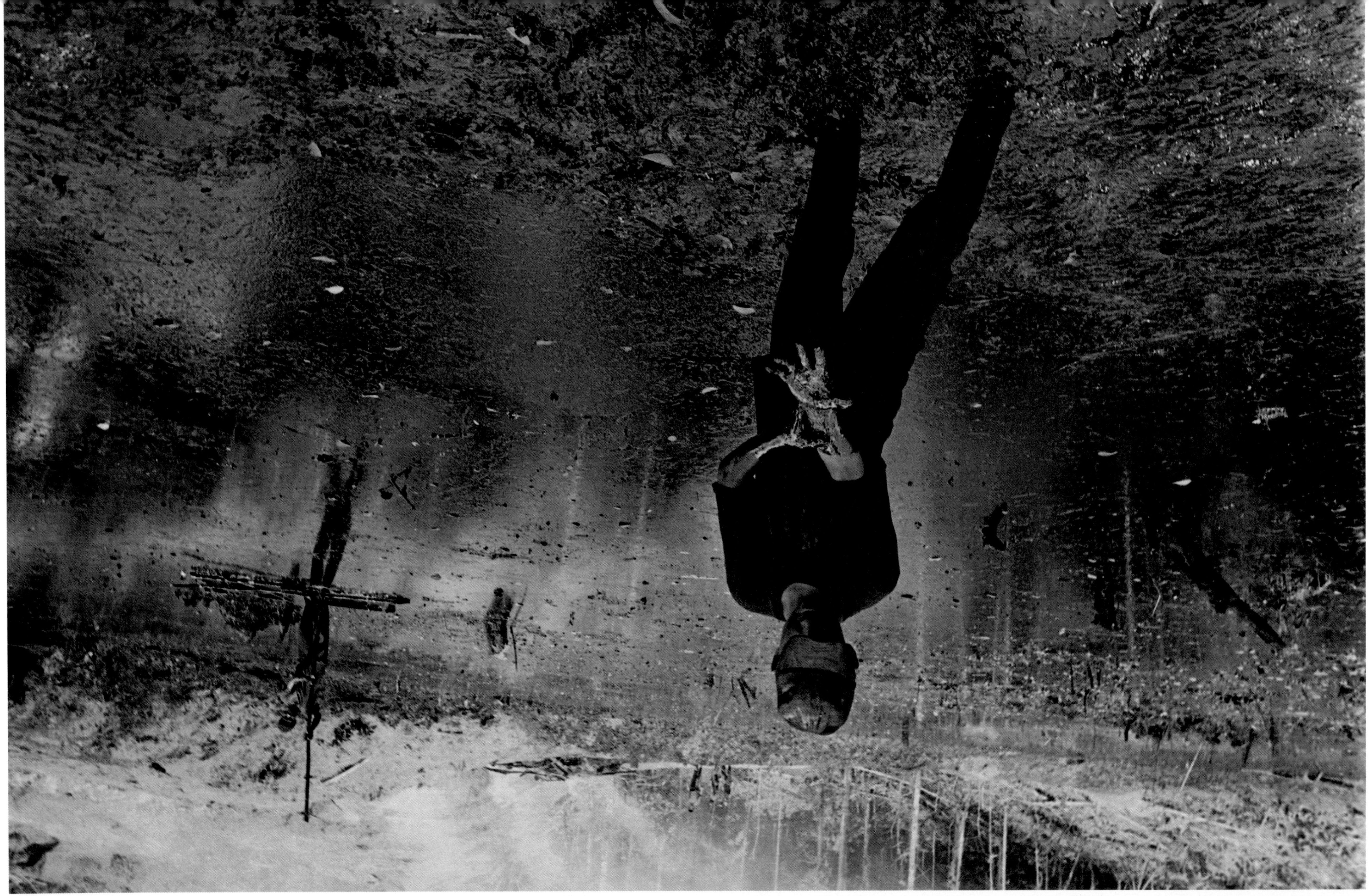

_ 119 WORKERS CLEANING UP AN OIL SPILL IN ORIENTE

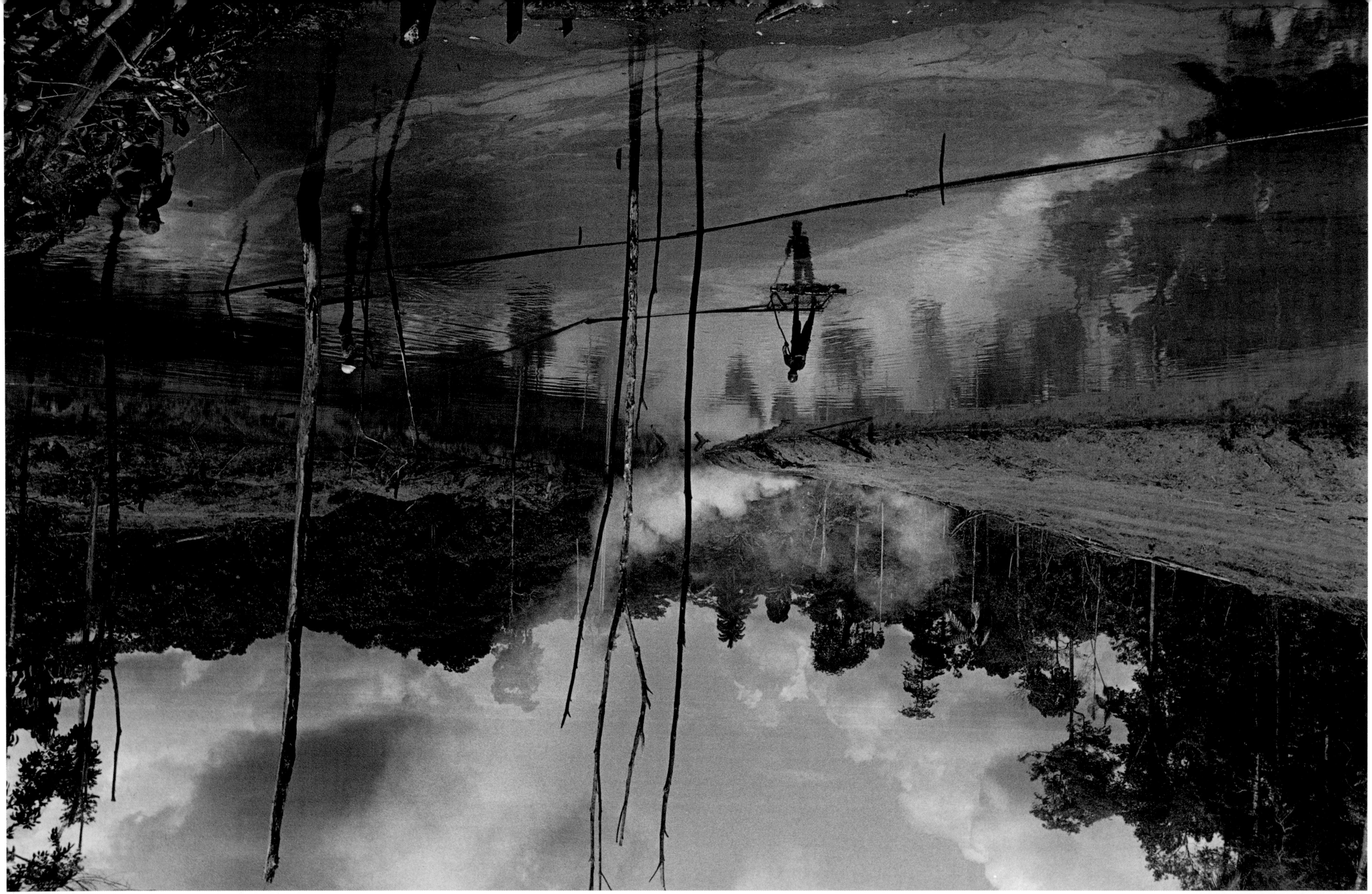

121 CONTAMINATED LAKE IN ECUADOR'S RAVAGED ORIENTE

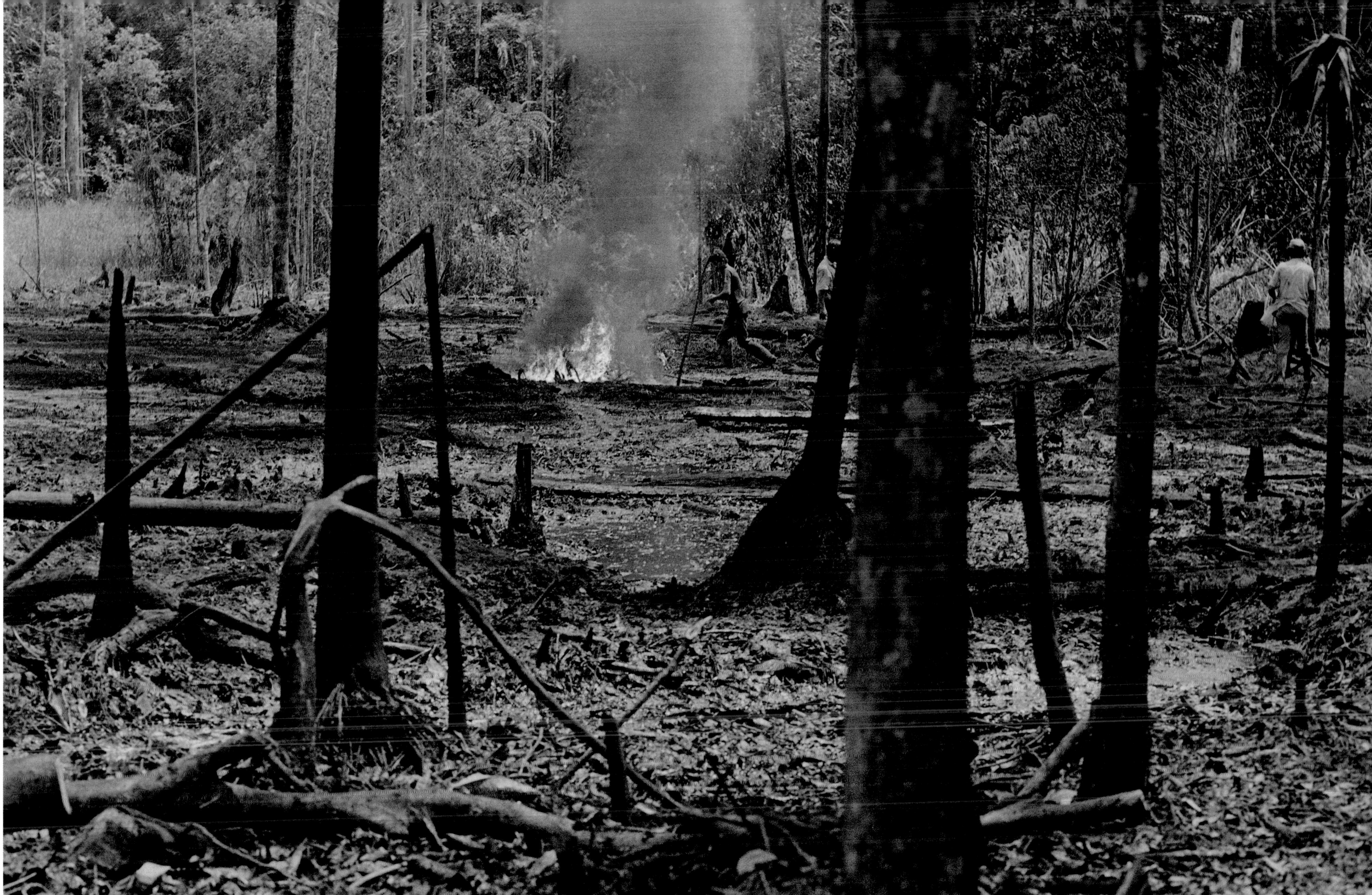

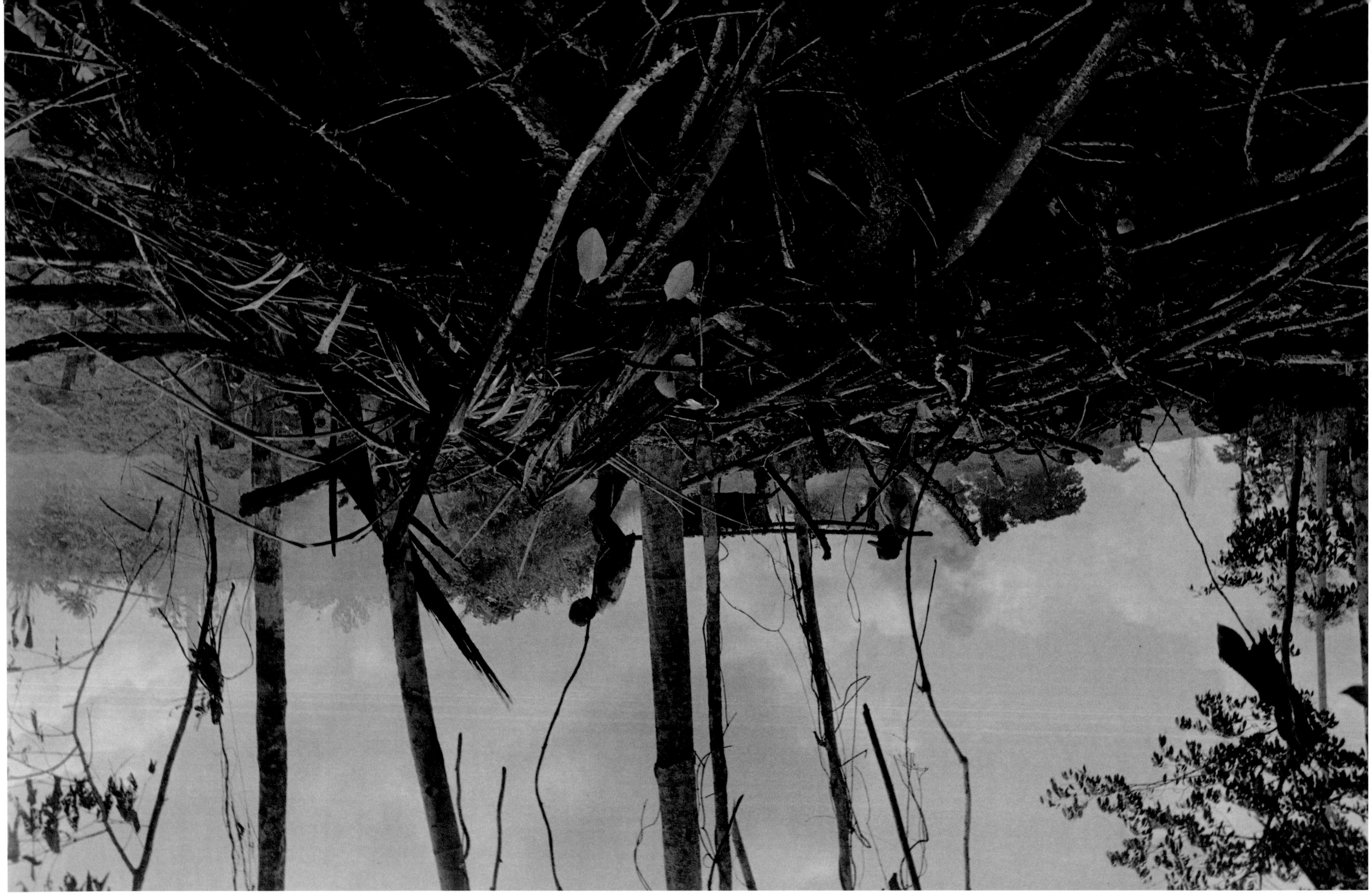

125 CLEANUP WORKERS CARRYING OIL TO AN OPEN DUMP

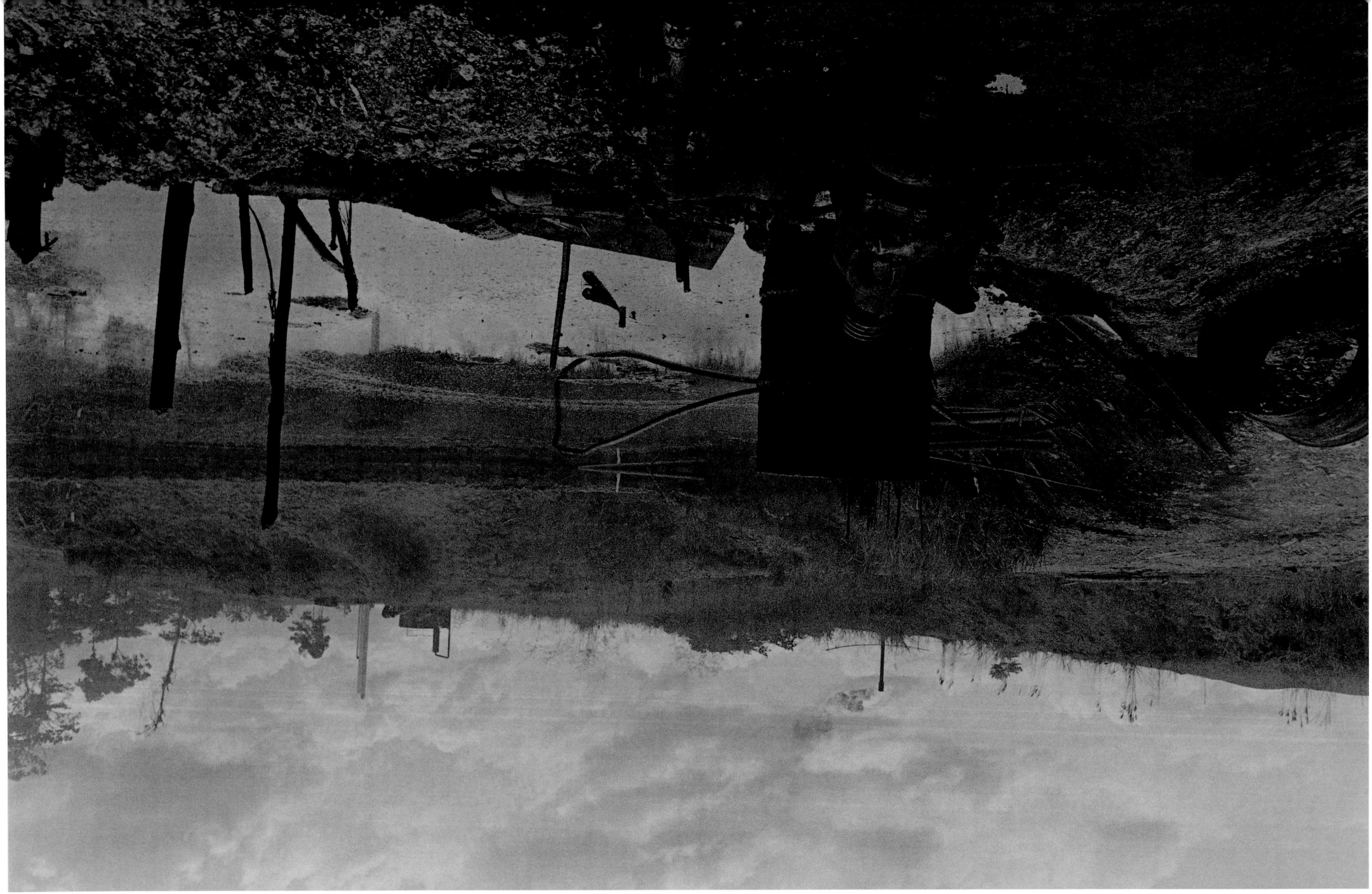

“ *I work for HSBC, the world's largest banking organization in Bangalore, India.*
I am a call center representative.
I have an Indian name but HSBC has given me a western one to use.
When people across the United States ask me who I am, I tell them Lindy.
We are briefed each day on American cultural facts —
television,
music,
film, and
sports.
And we are instructed to use these topical details when talking with customers.
It makes them feel comfortable.
A speech instructor has neutralized my accent so that anyone listening thinks I'm from where they're from.
Recently, I had a customer tell me about the weather in Phoenix, Arizona.
I live in Bangalore and have an accent that comes from nowhere, and I've never been to America. ”

A new dynamic has come about in the business world,
one that distances production and service from those who control and price the product.
The burgeoning call centers throughout India are the testaments to this brave new creation.
In Bangalore and Hyderabad,
Indian college graduates man the telephones of multi-national corporations
answering calls from customers in Europe and the United States.
Banking,
computer-technical support and
credit card companies
now employ workers who come cheaper than those in the countries
where the products or services are sold.
A call center operator in the United States earns $30,000 per year.
In India it's $5,000.

THE ERA OF [PEOPLE AS INTERCHANGEABLE COMPONENTS] IN THE SYSTEM OF MANUFACTURE IS HERE.

HUMANS ARE CAPITAL.

prague _ homoglobalus

_ *Section*

IX

HOMOGLOBALUS

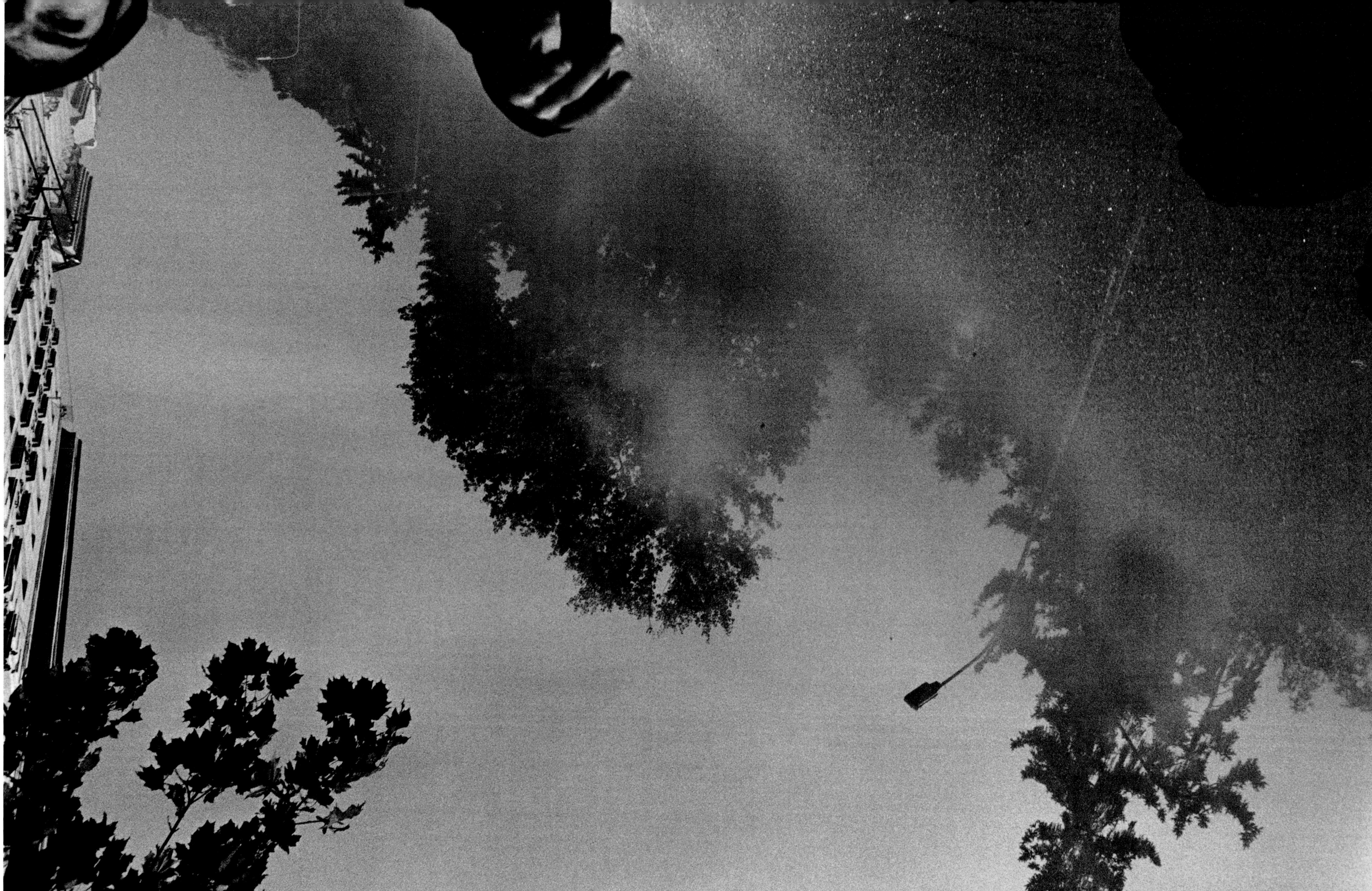

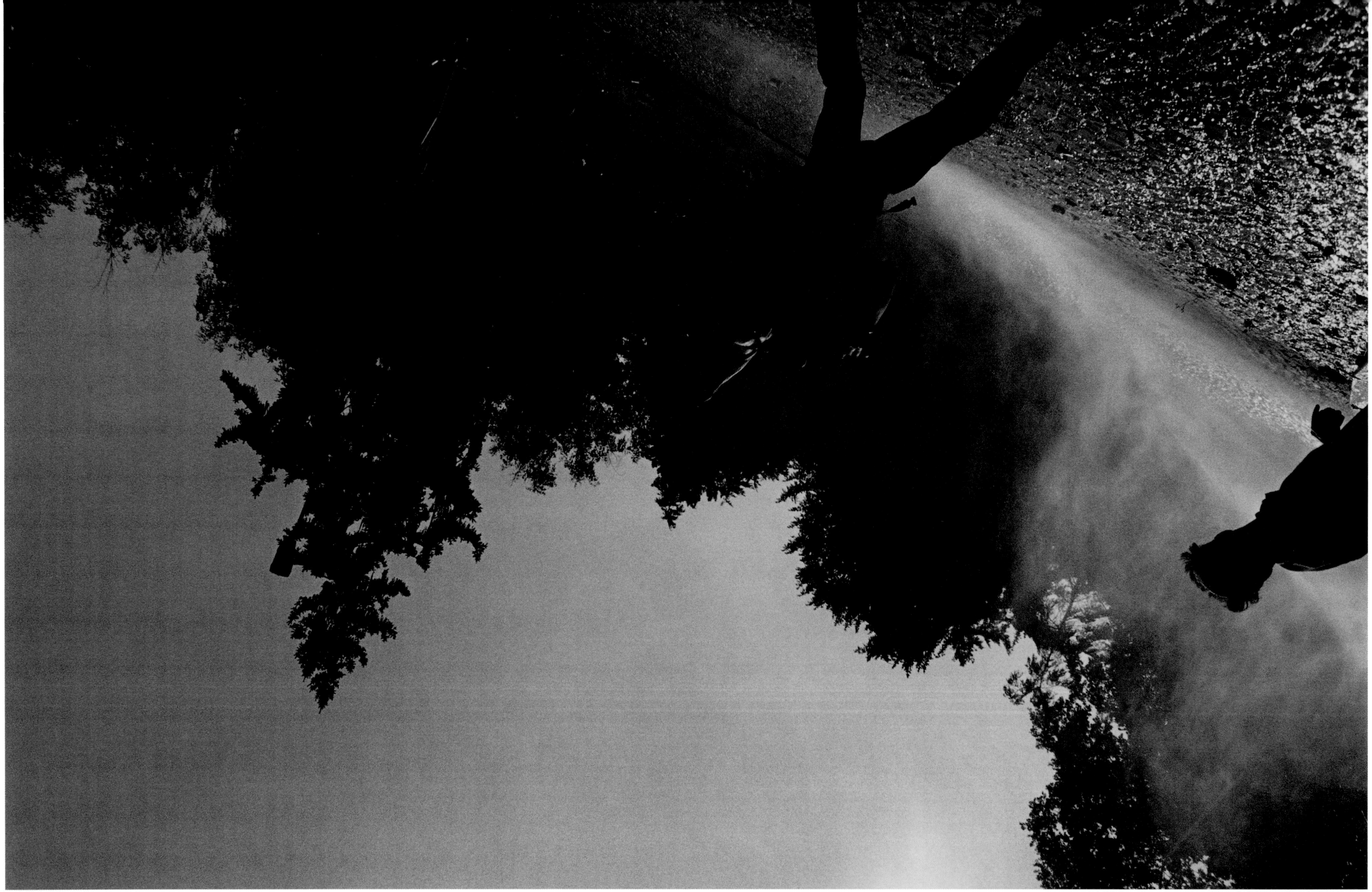

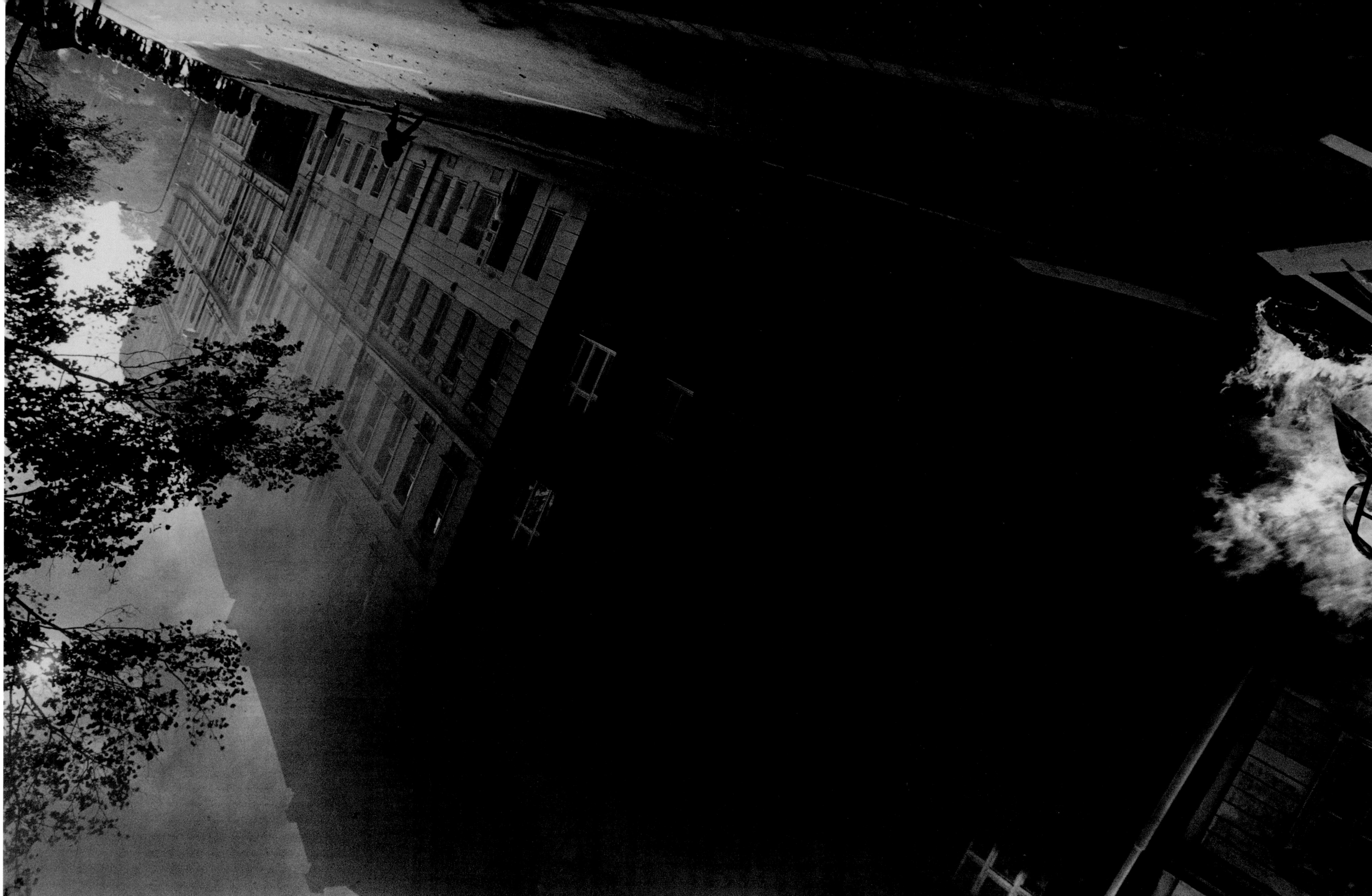

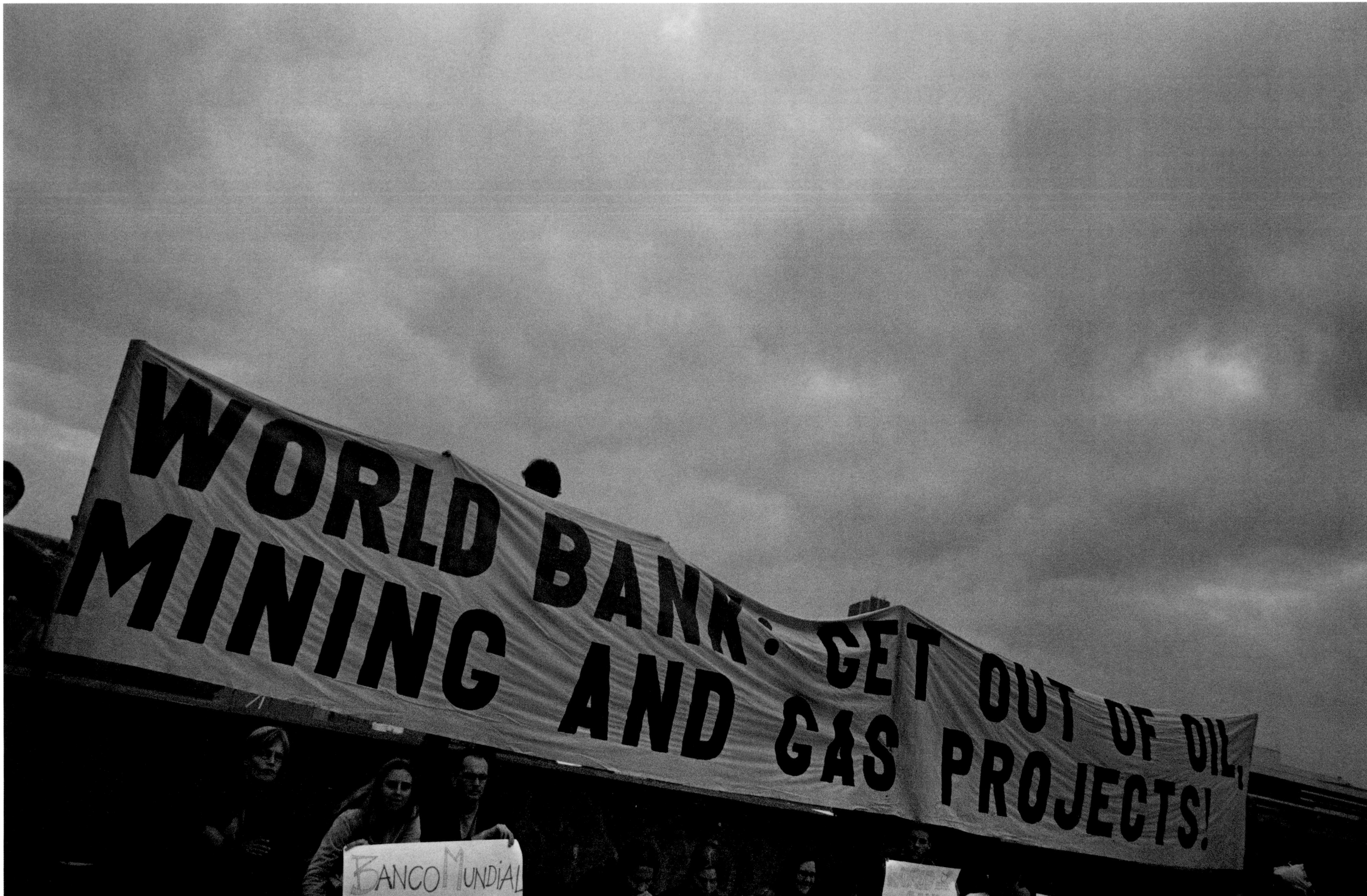
WORLD BANK: GET OUT OF OIL, MINING AND GAS PROJECTS!
BANCO MUNDIAL

PROTESTERS HOLD A BANNER OUTSIDE THE CONFERENCE

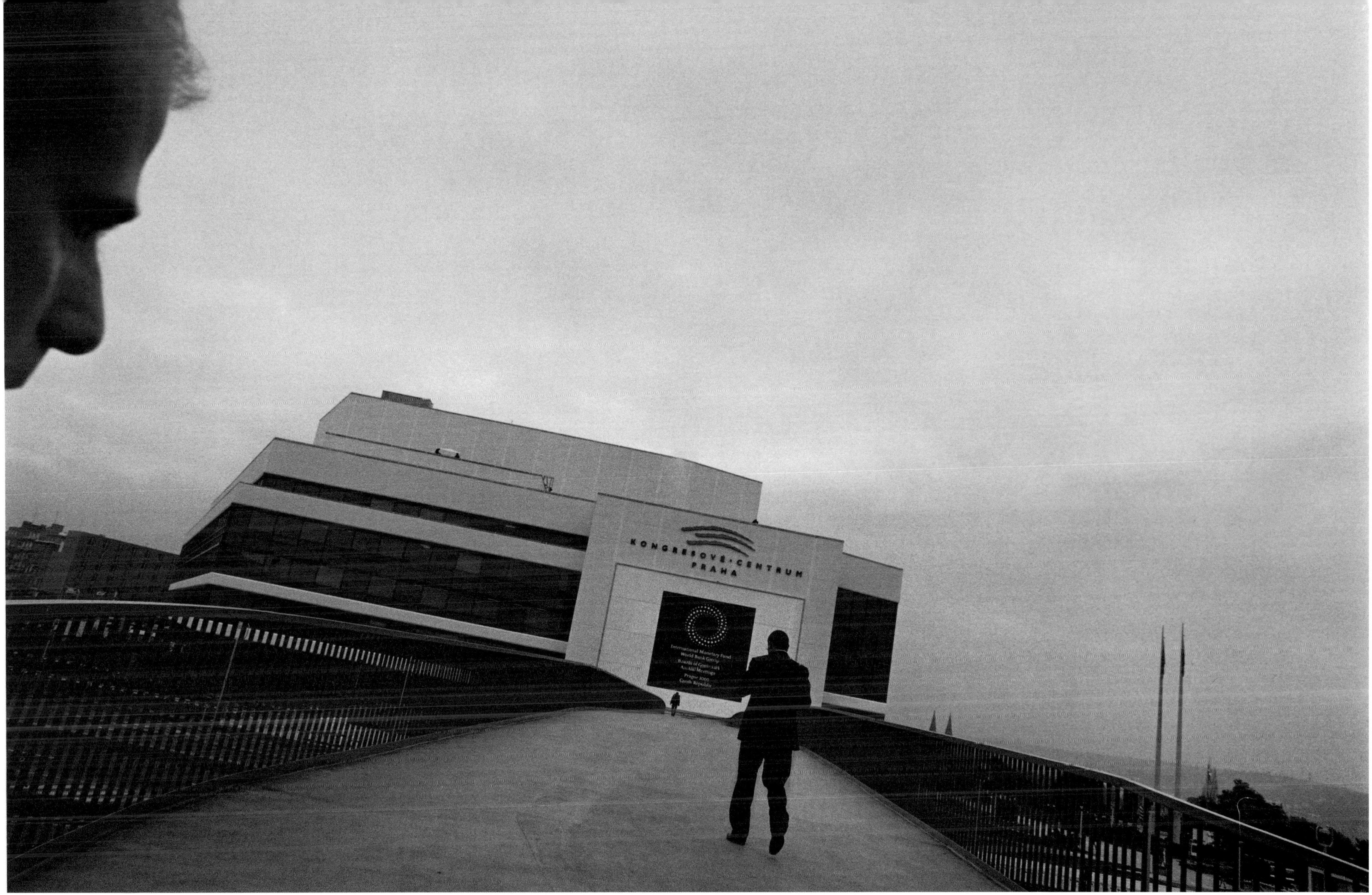
KONGRESOVÉ·CENTRUM
PRAHA
International Monetary Fund
World Bank Group
Prague 2000

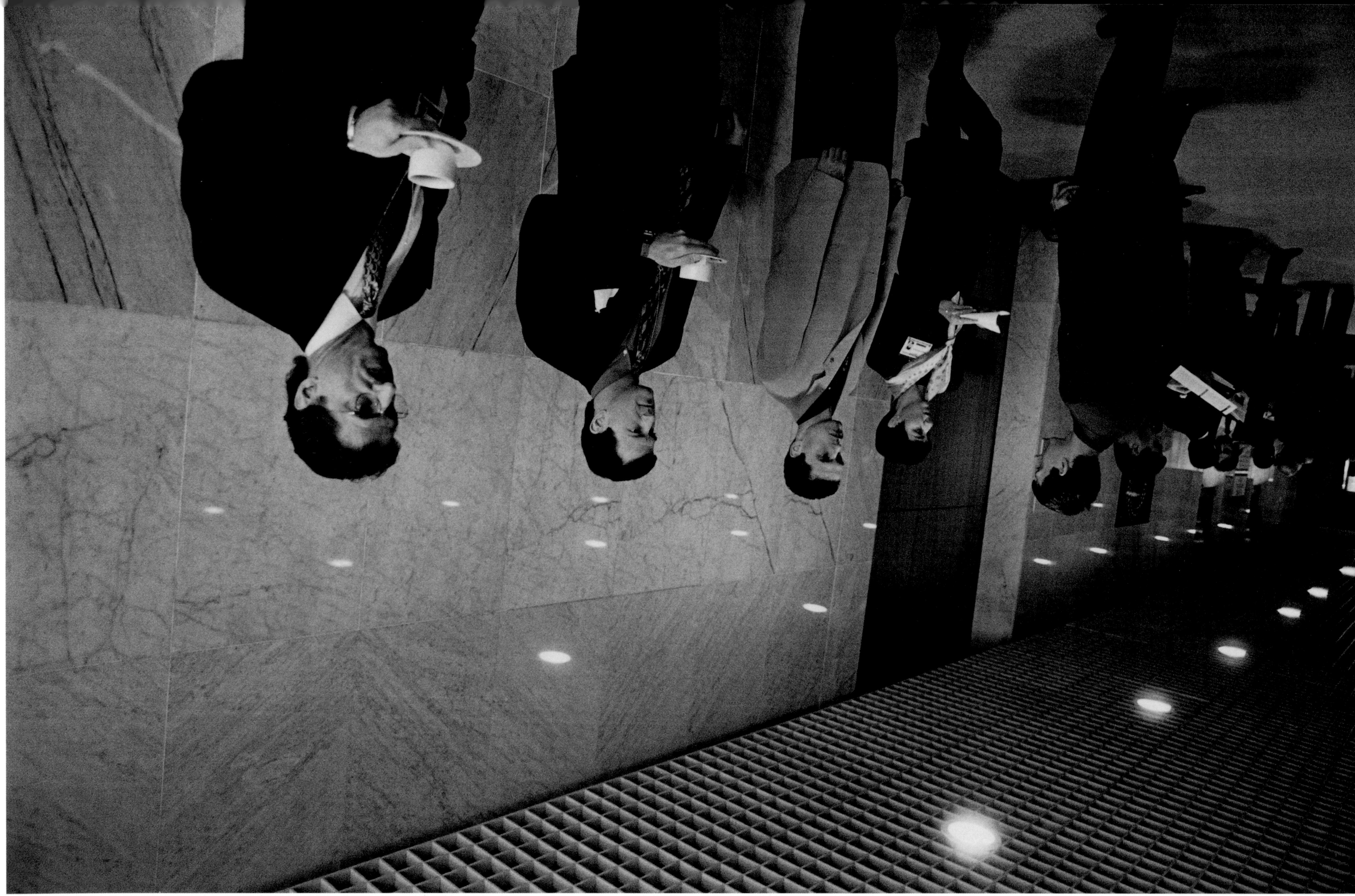

“ In 2002, I spent a month in Belareus.
There, I traveled to the region known as GOMELSKAYA and spent a week observing this cut of land that received.
more radioactive fall out from Chernobyl's nuclear catastrophe than anywhere else.
In 2002 the average level of radioactivity in Gomelskaya exceeded the norm by 200 times.
It had been over a decade and a half since the disaster took place.

EACH DAY WE WERE GIVEN PERMISSION TO ENTER [THE ZONE] .

[THE ZONE] **is the closed strip along the power plant's perimeter, in Ukraine, flanking the borders of Belarus.**
It was forbidden to stay there for more than eight hours at a time.
And to this restriction was added no eating of anything and drinking only water from closed bottles.

Entering the [THE ZONE] , whatever preconceptions I might have had of a horribly mutated post-apocalyptic landscape immediately disappeared.
Here was something altogether different.

With man's influence having been removed entirely for the last 16 years, nature had not only prospered,
it had taken back what had been built on Nature had devoured human civilization.
Trees shot through houses, their limbs lifting the very roofs off the structures themselves.
A VILLAGE SANK INTO WILD VEGETATION.
Rivers, once again, flowed along their natural courses finding their original beds.
Chairs sat in trees several meters off the ground carried up simply for having had something sprout beneath.
In the middle of a road lay a gynecological chair.
It had been de-fenestrated by a wild bush that now occupied a doctor's practice entirely.
Everything appeared healthy and strong.
Nature's convincing return showed that man was in fact its most dangerous threat.
It was clear that were man to leave nature alone, nothing could endanger it, not even a nuclear disaster.

The Belarusian government had begun implementing a limited agricultural program in the outer zone.
Whilst there, I saw tractors plough the contaminated fields.
The government's logic was to grow crops only for animal feed.
The radioactive feed that would be eaten by cows would contain radioactive Iodine 131 and.
Strontium that would deposit themselves in their bones;
however, the third material, Cesium 137, would dissipate from the animal after 90 days, resulting in meat fit for human consumption.
Using bones for soup, though, was forbidden. Bones were contaminated. They were poisoned.

What farmers would most complain about was the dust that blew around the fields whilst ploughing.
It is a fine dust from the surface of the soil and is the most contaminated element that sits on the ground and hangs in the air once it's stirred.
Tractor cabins are now in the process of being hermetically sealed to prevent further inhalation of this radioactive substance.

We had searched for people who had returned to the restricted area.
Eventually, we found some: an old grandmother and grandfather who lived in one of the evacuated villages.
“Well, that government of ours makes a big deal out of it all!”
said the old man.
“Nothing smells, things don't taste bad, you can't even see it ... so how can it be bad?”
A young scientist accompanying us measured the radiation in the old couple's garden.
It was 160 times higher than anywhere else he'd been that day.
The couple ate the vegetables from their garden, drank the milk from their cow and the old man smoked the bark off the trees which grew all.
around. They were over 70, they said, and experienced no pain and felt extremely happy.
We gave the old man $20 to run to town and buy vodka.

Following the nuclear inferno the Soviet Government handed out double rations of vodka to those living closest to the burning power plant.

We'd had a guide helping us during our stay in [THE ZONE] . He would interpret and knew his way around the place.
At the time of the explosion he had been a conscript in the military and as a result
been assigned the task of transporting the contaminated soil from the worst affected areas of the country.
His assingment had taken one month.
HE WAS 35 BUT LOOKED 50. And had heart problems.
Out of the 40 conscripts working along side our guide in those first months, less than half were alive.
He had said pinning his colleagues deaths as well as his health problems to radioactivity wasn't easy.
ONE COLLEAGUE HAD DIED FROM A HEART ATTACK , ANOTHER A STROKE . ONE HAD LIVER FAILURE , ANOTHER CANCER .
BUT NONE OF THEM DIED EXPLICITLY THROUGH RADIATION .

If you didn't know about the staggering incidence of cancer in this place,
of the abnormally high infertility here,
that the women from Gomelskaya had to lie about where they'd fled from so as not to lose their partners ...
who wants a woman who in all likelihood will only bear handicapped children?
If I hadn't heard the stories from these radioactive refugees telling of the abuse they suffered by being made to feel like
lepers from people living in other parts.
If I didn't know this, then [THE ZONE] was a paradise where nature had won a great victory over mankind.
Where trees, bushes and grass had smothered everything man-made for centuries yet to come.
If I didn't know what man had suffered because of himself,
I'd have to say he'd deserved it. "

Marek Dusak — Czech documentary filmmaker

chernobyl _ a nuclear eden

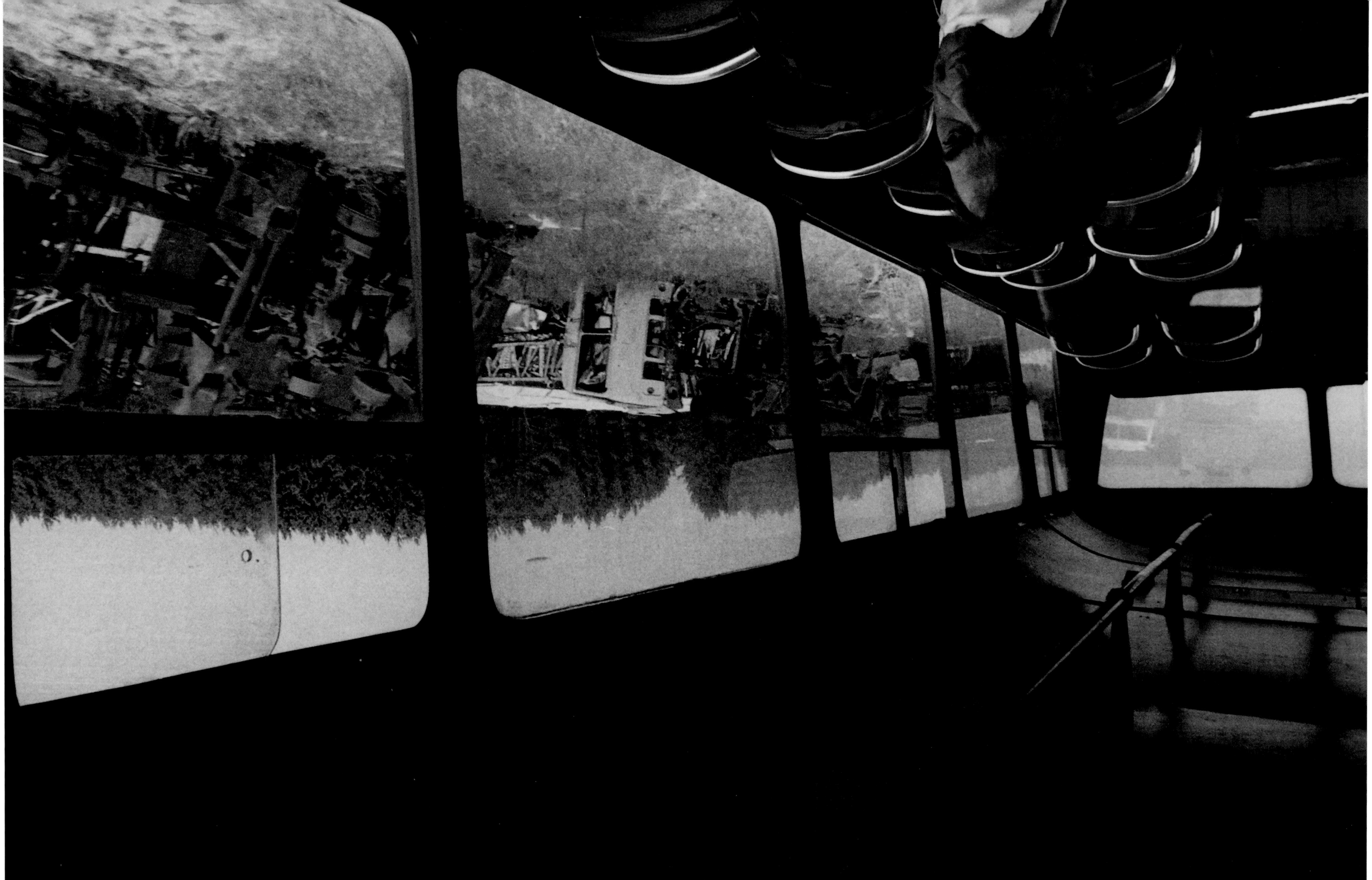

_ 145 CLEANUP ENGINEER HEADING TOWARDS RADIOACTIVE CEMETERY
(BURIAL GROUND FOR CONTAMINATED TANKS AND HEAVY EQUIPMENT)

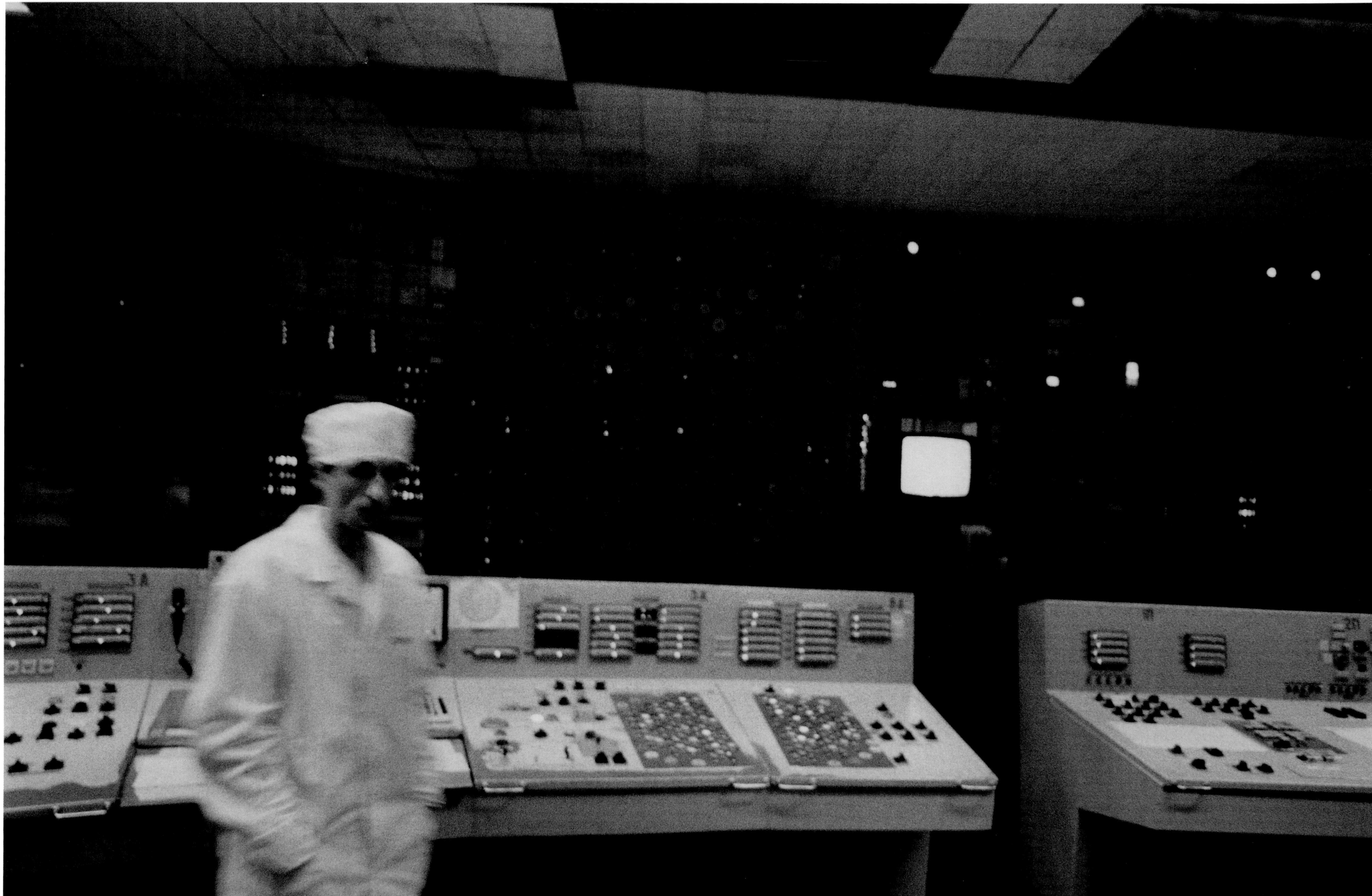

_ 147 CONTOL ROOM, CHERNOBYL REACTOR NO. 3
_ 148 MAN DYING OF CANCER IN A VILLAGE BORDERING THE FORBIDDEN ZONE
_ 149 BOY WITH RADIOACTIVE MILK

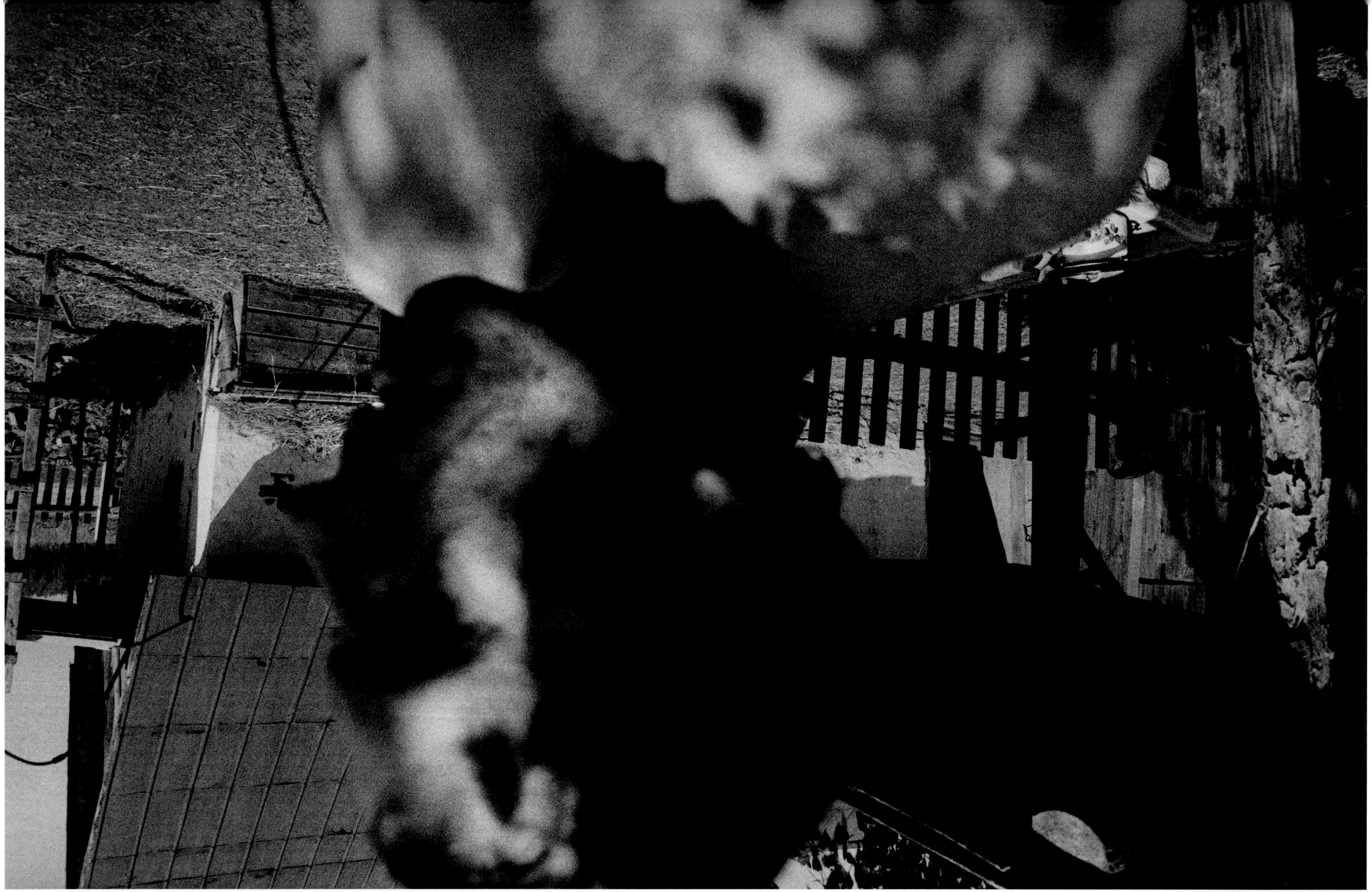

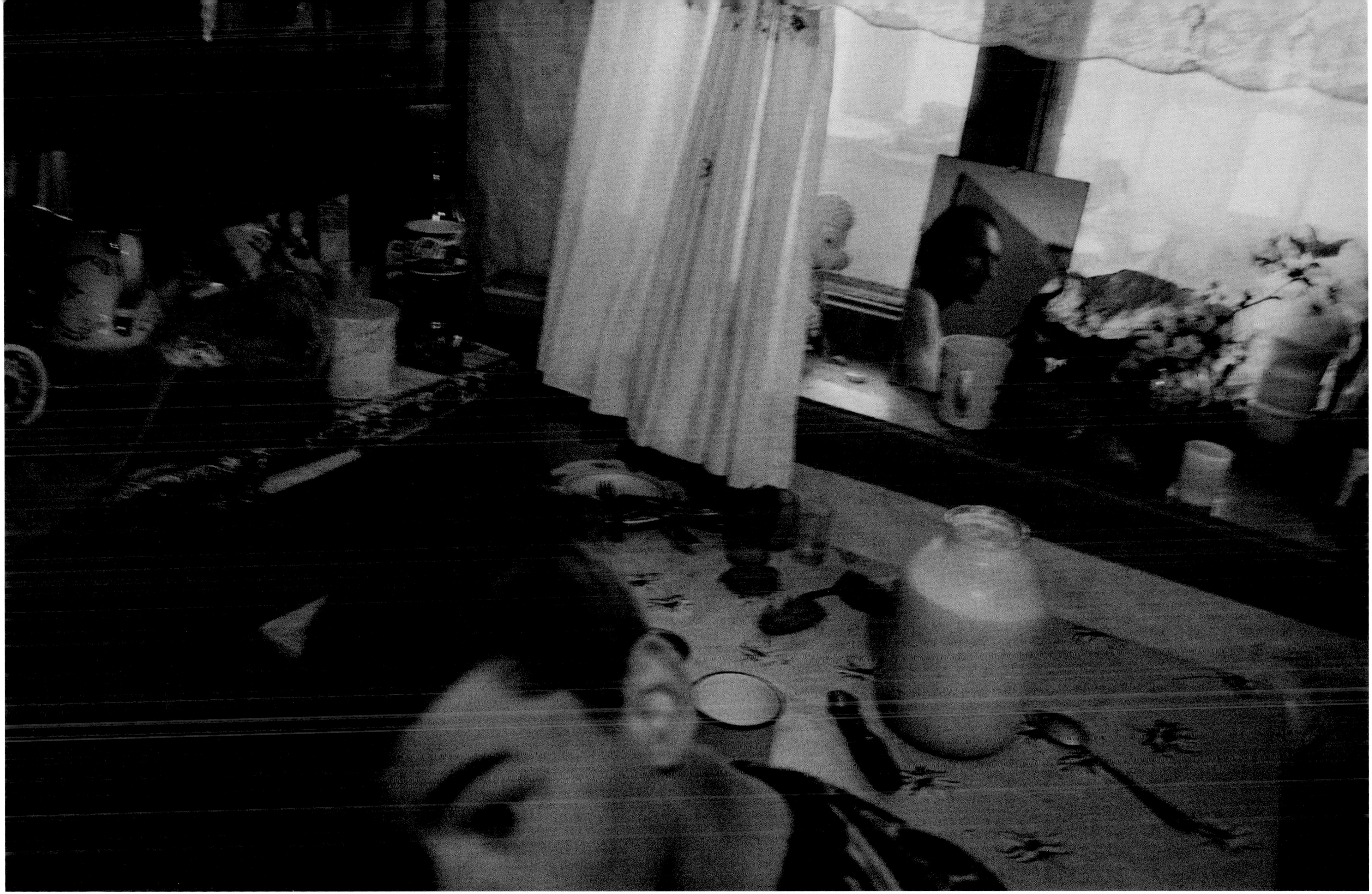

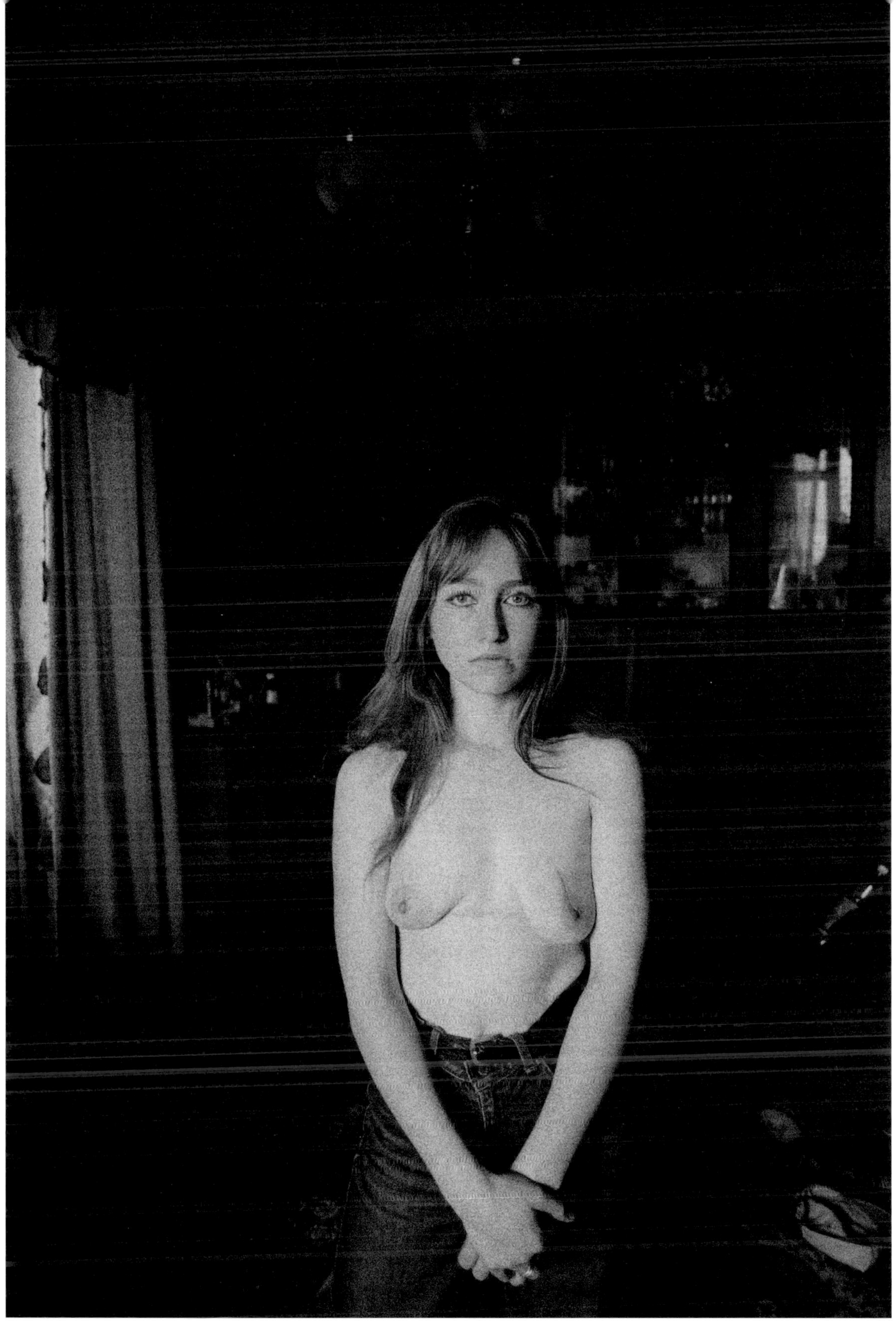

_ 151 THE EFFECTS OF RADIOACTIVE EXPOSURE
_ 152 AMUSEMENT PARK IN THE ABANDONED CITY OF PRYPIAT
_ 153 ABANDONED APARTMENT. PRYPIAT

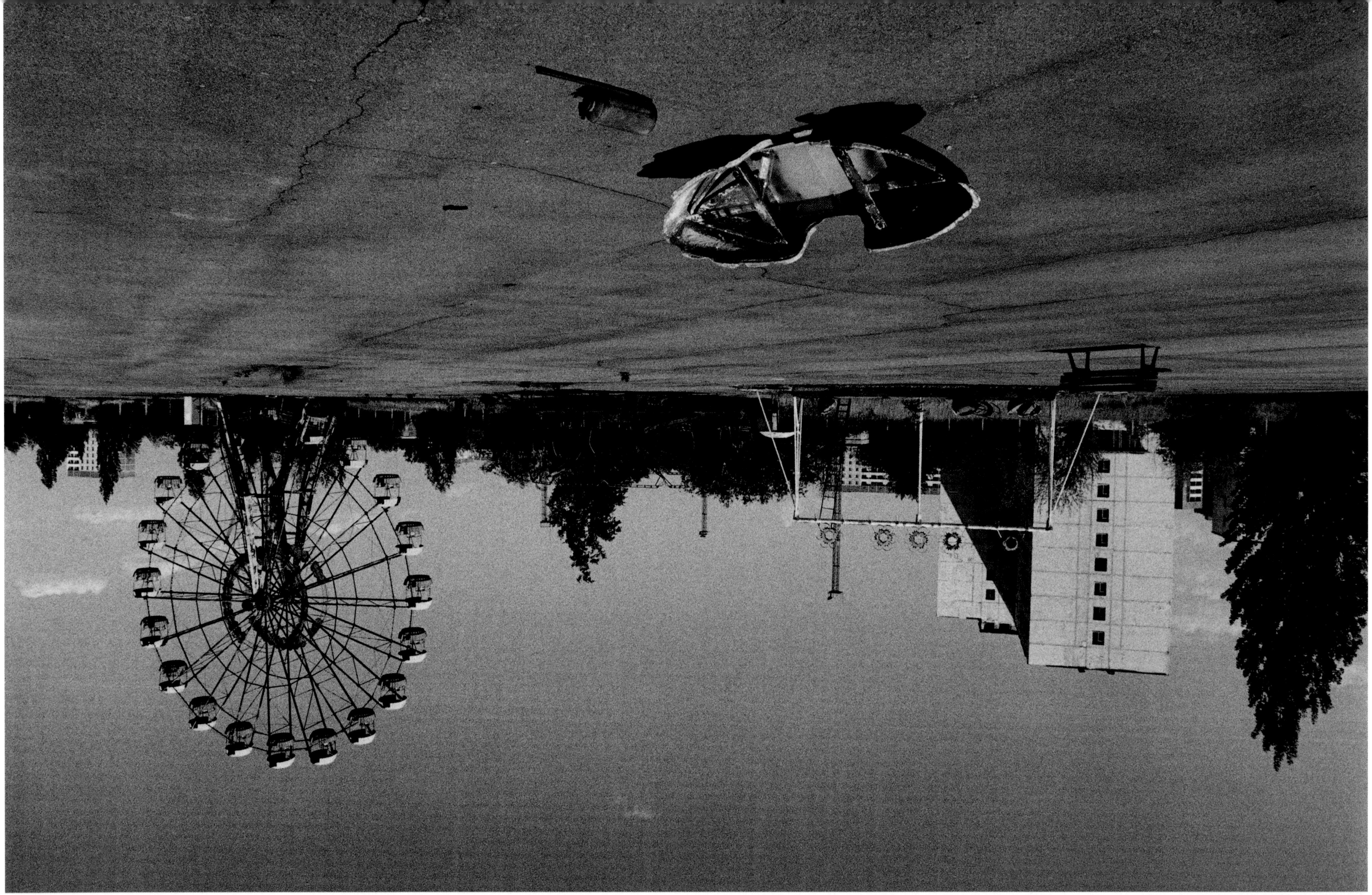

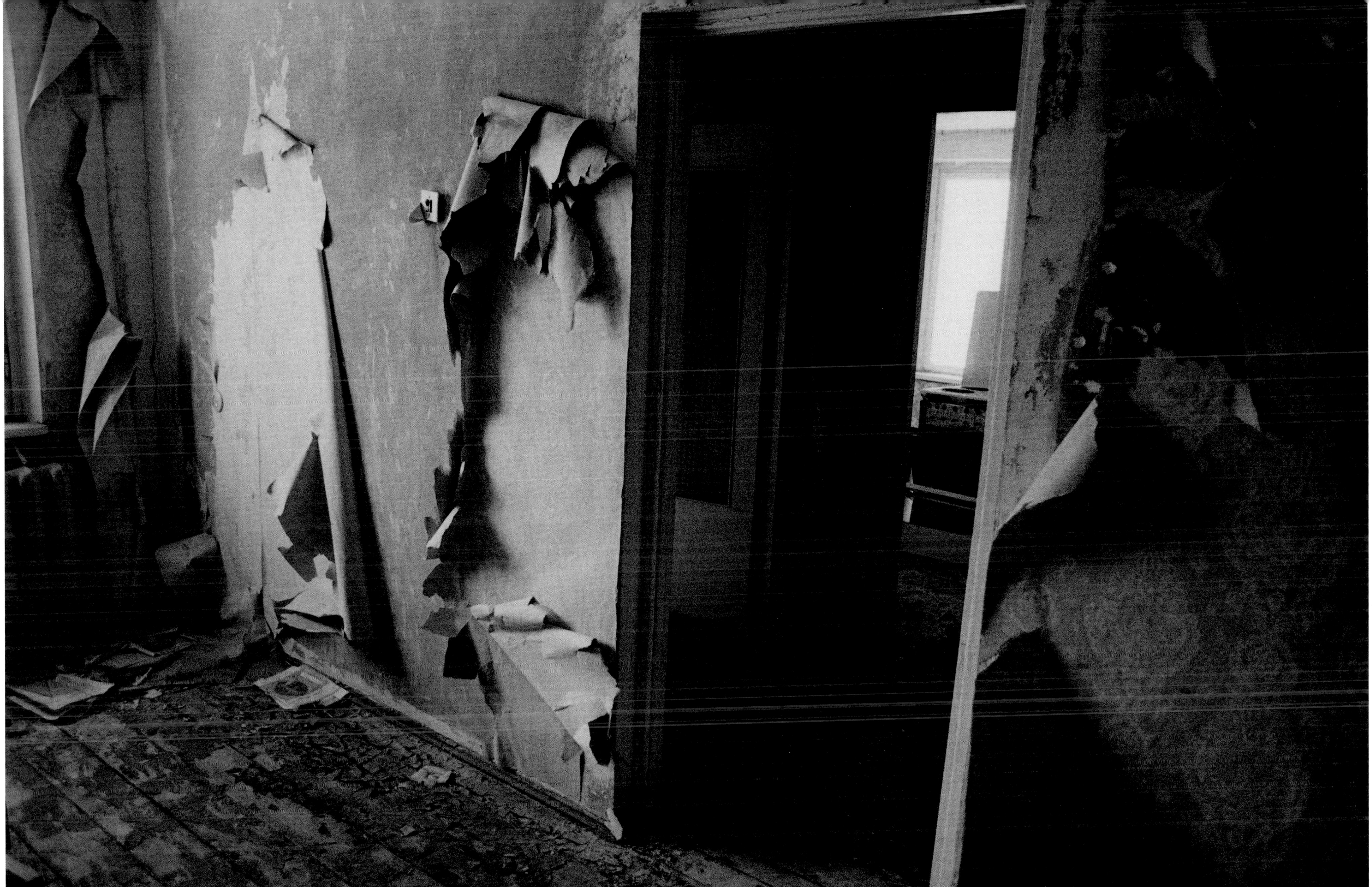

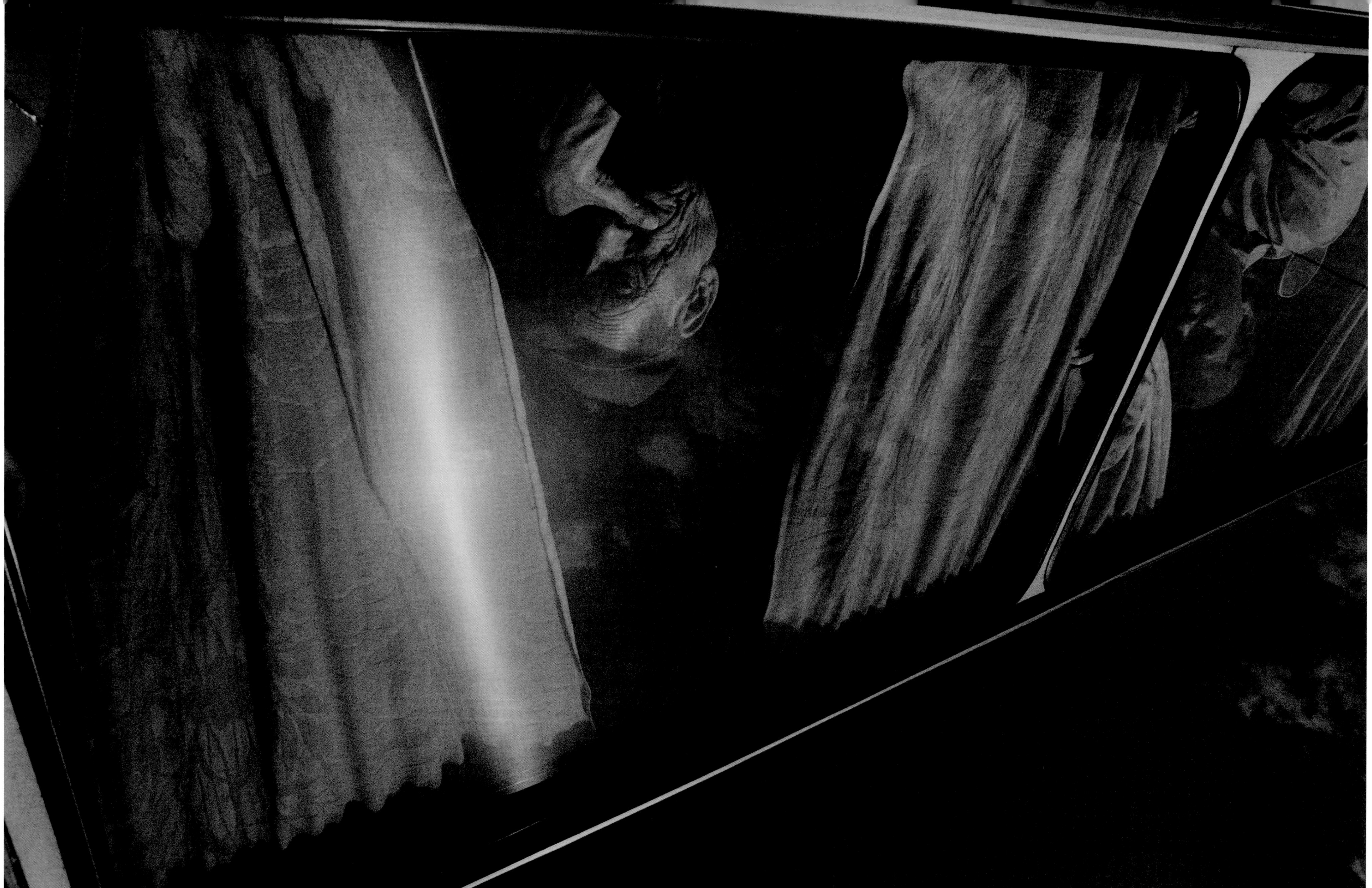

_ 155 FORMER RESIDENTS OF THE RADIOACTIVE TOWN OF PRYPIAT ARE GIVEN ONE HOUR TO VISIT THEIR ABANDONED HOMES

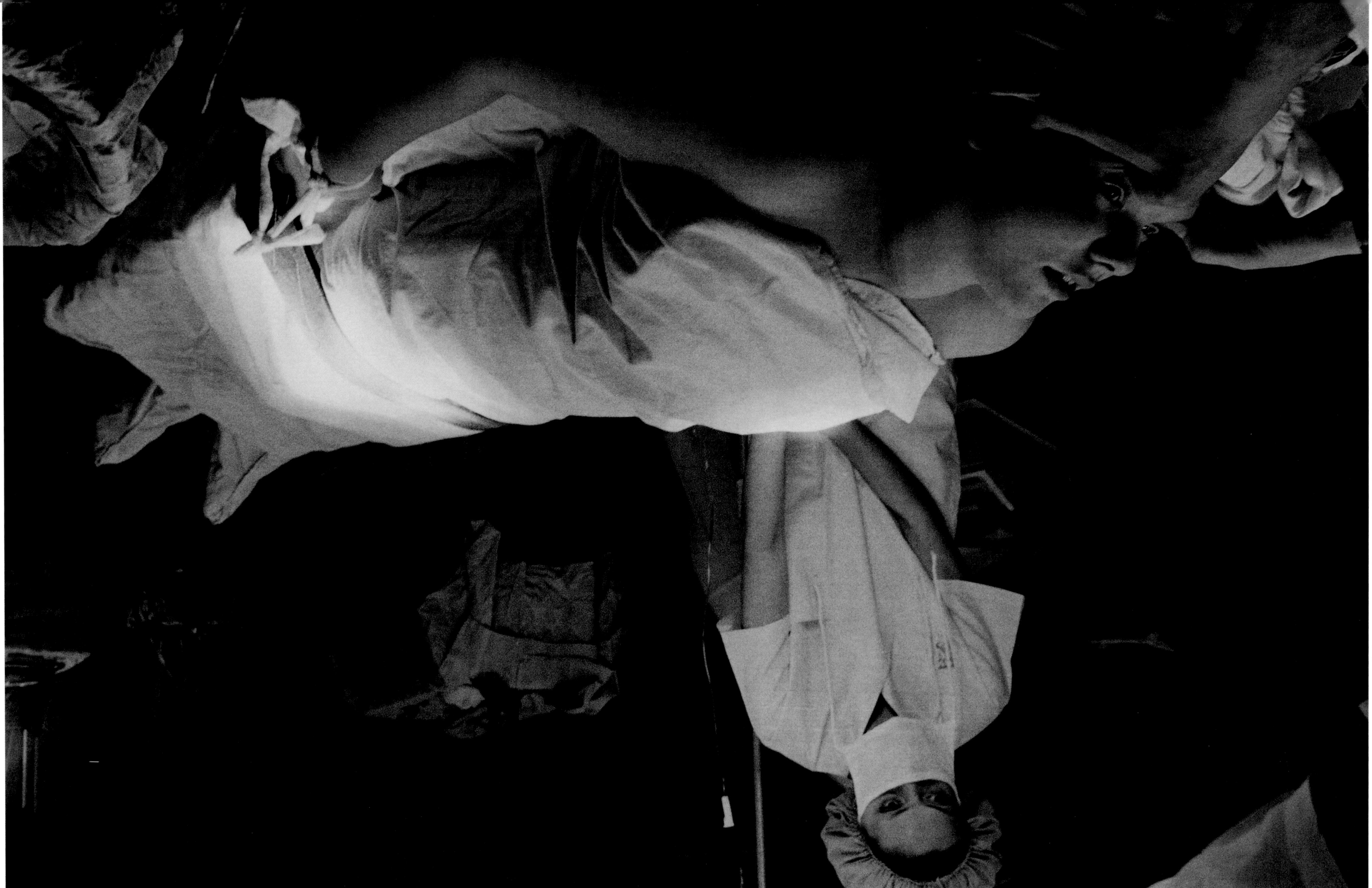

SOME SAY YOU CAN'T GET BLOOD OUT OF A STONE.
ANGOLA'S DIAMONDS SAY DIFFERENT.

Jonas Savimbi, the leader of Angola's once American-backed UNITA rebel forces,
fought a quarter century long civil war with the profits made from selling these gems.

THESE WERE THE MINERAL WARS.

At one time UNITA's illegal diamond trade was estimated at $1,000,000 a day.
The stones bought thousands of guns, scores of tanks and millions of anti-personnel mines.
Mines are the perfect soldier — never sleeping, never off-target, and, like diamonds became the gifts that kept on giving.
Seventy thousand Angolans alone were maimed by land mines.
Those killed by them double that.

It takes between 11 to 35 pounds (5 to 16kg) of pressure.
to detonate a mine.
The initial blast tears the foot apart,
causing it and its toes to peel away from the leg.
The blast forces dirt,
bits of shoe,
mine fragments,
bone and tissue.
to be driven deep up.
into what remains.
of the leg,
the genitalia and torso.

The outer skin returns into place and hides.
the full extent of the damage.
Victims rarely make it out of the mine field alive.
They die in agony where they lie.

OVER HALF ARE WOMEN AND CHILDREN.

A victim remembers,
"I lay there.
I was bleeding everywhere.
I had blood in my mouth.
I looked down and saw my foot next to my leg.
I reached down and tried putting it back on but it kept falling off.
I tried again and again. It wouldn't stay."

In 1986 SAVIMBI visited the White House.
President Ronald Reagan described him as AFRICA'S GREATEST FREEDOM-FIGHTER.
What diamond customers around the world were buying in.
dollars, pounds, francs and Dutch guilders.
had already been paid for.
with the amputated limbs and lives of Angolans.

and their insurrections against internationally recognized governments. ”

President
George W. Bush
April 26th,
2004,
upon signing new legislation
to curb the
illegal trade in
blood
diamonds

XI :

angola _ bloody diamonds

_ *Section*

XI

BLOODY DIAMONDS

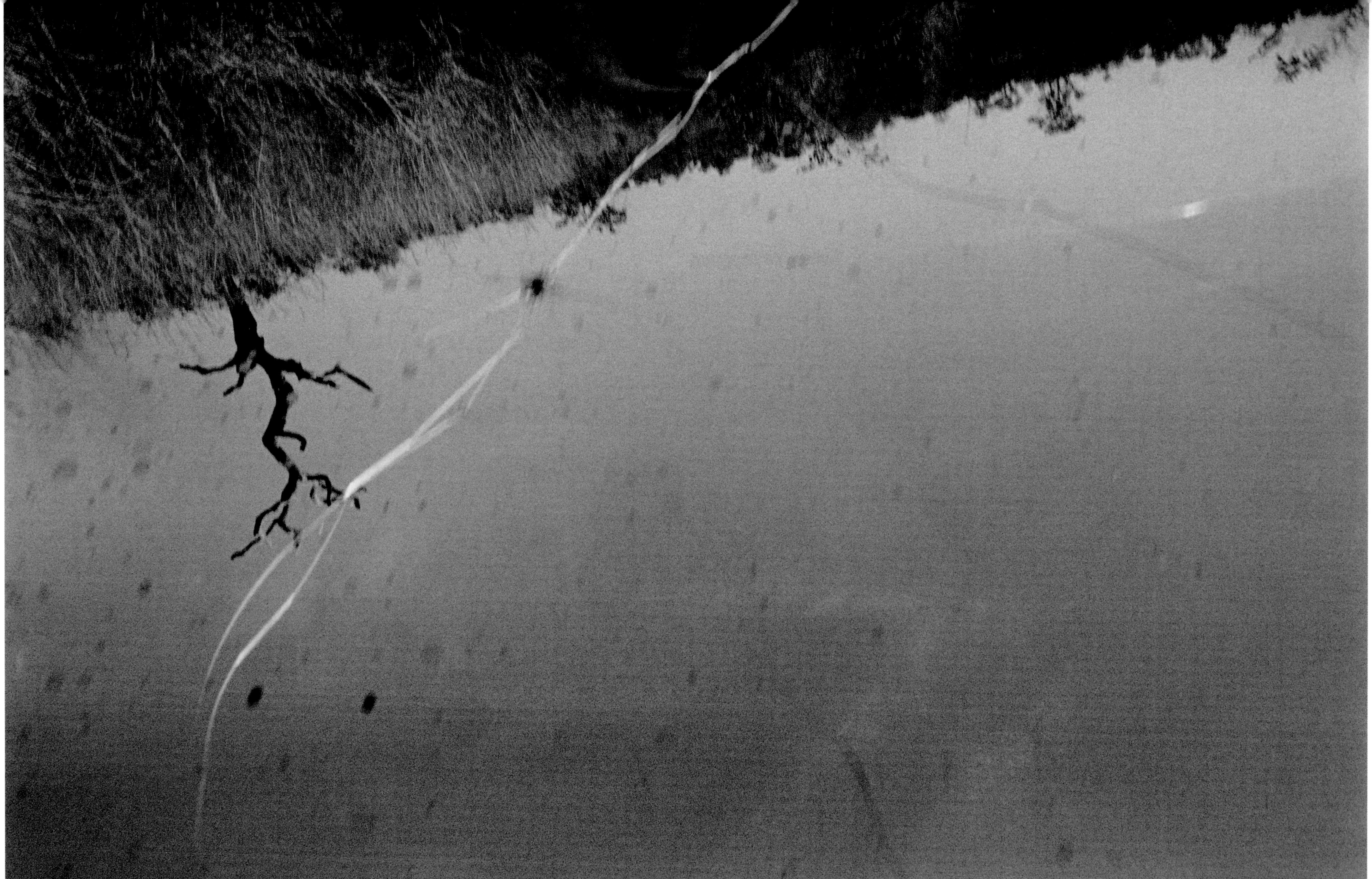

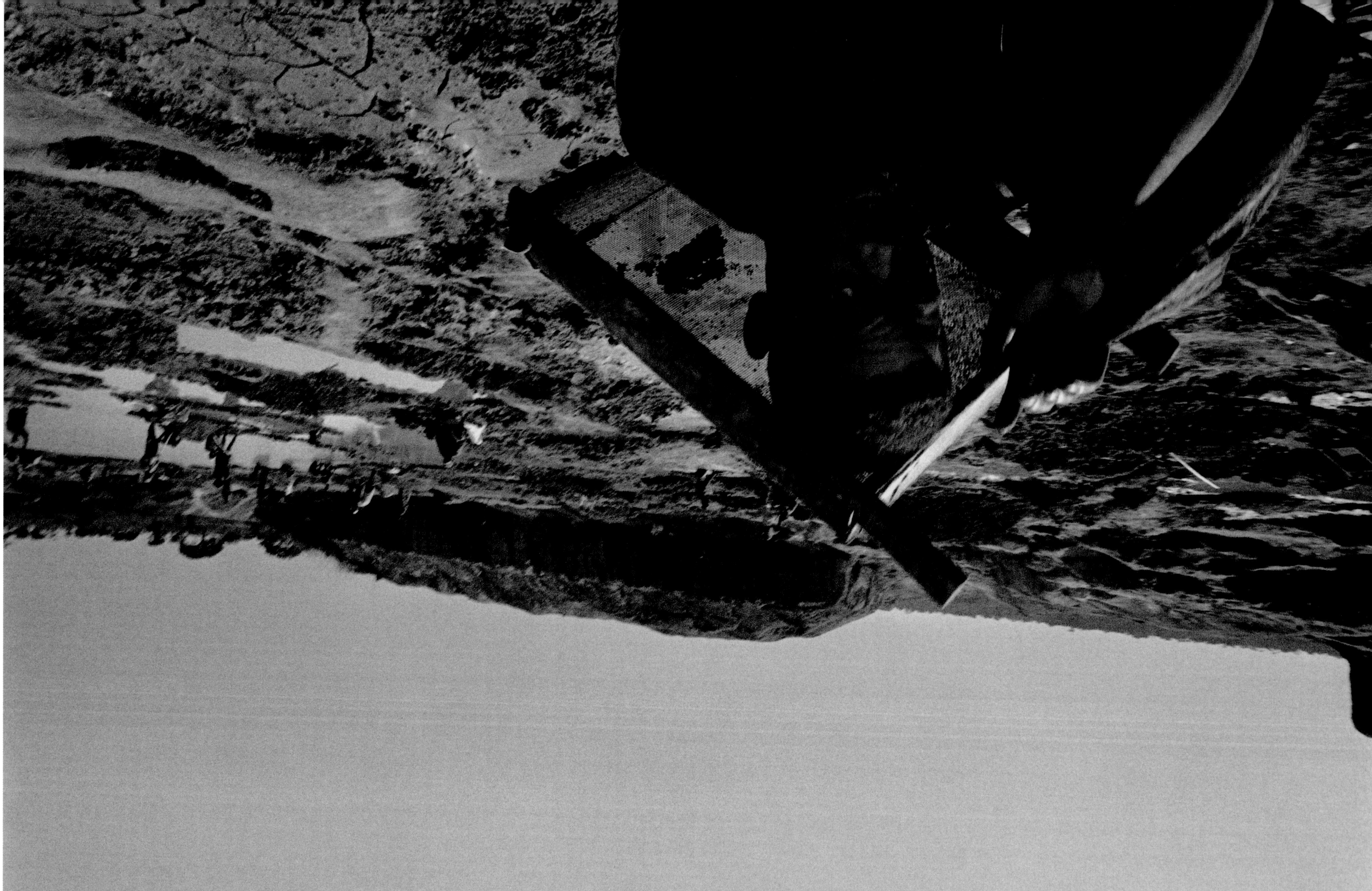

_ 163 ILLEGAL MINER CARRIES HIS SIEVE . UNITA TERRITORY
_ 164 HUNDREDS OF ILLEGAL DIAMOND MINERS . UNITA TERRITORY
_ 165 ILLEGAL DIAMOND MINE . UNITA TERRITORY

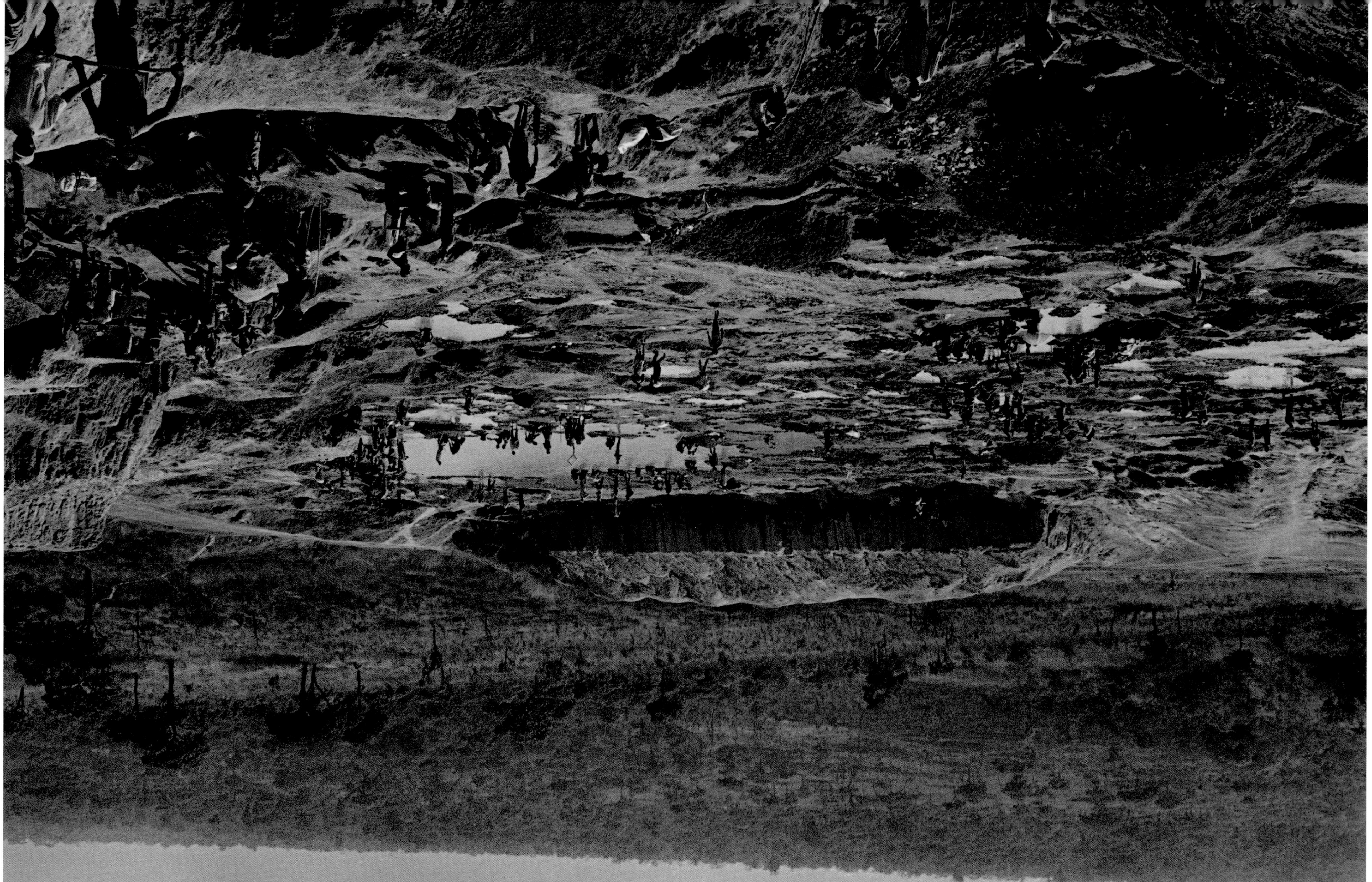

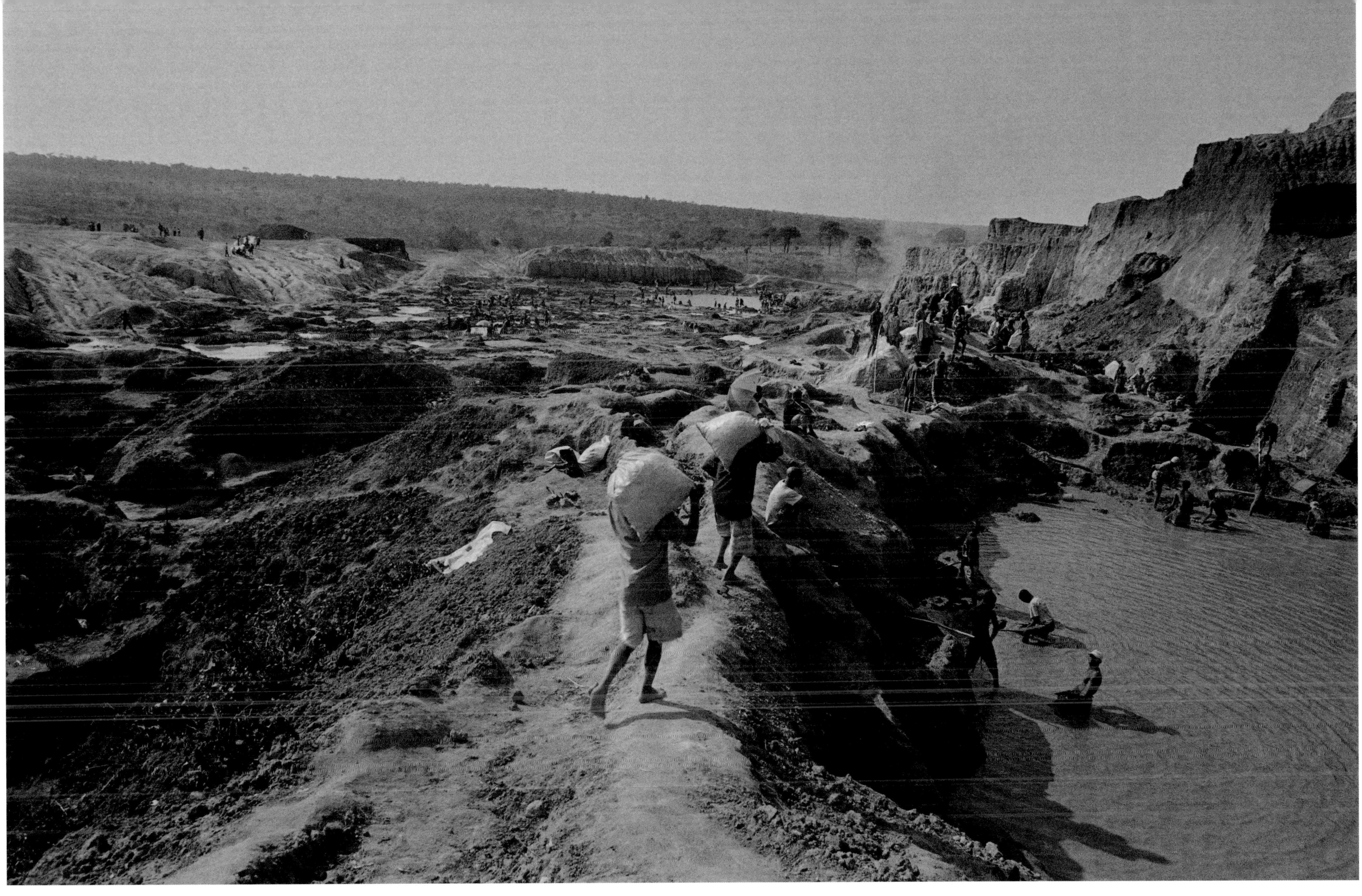

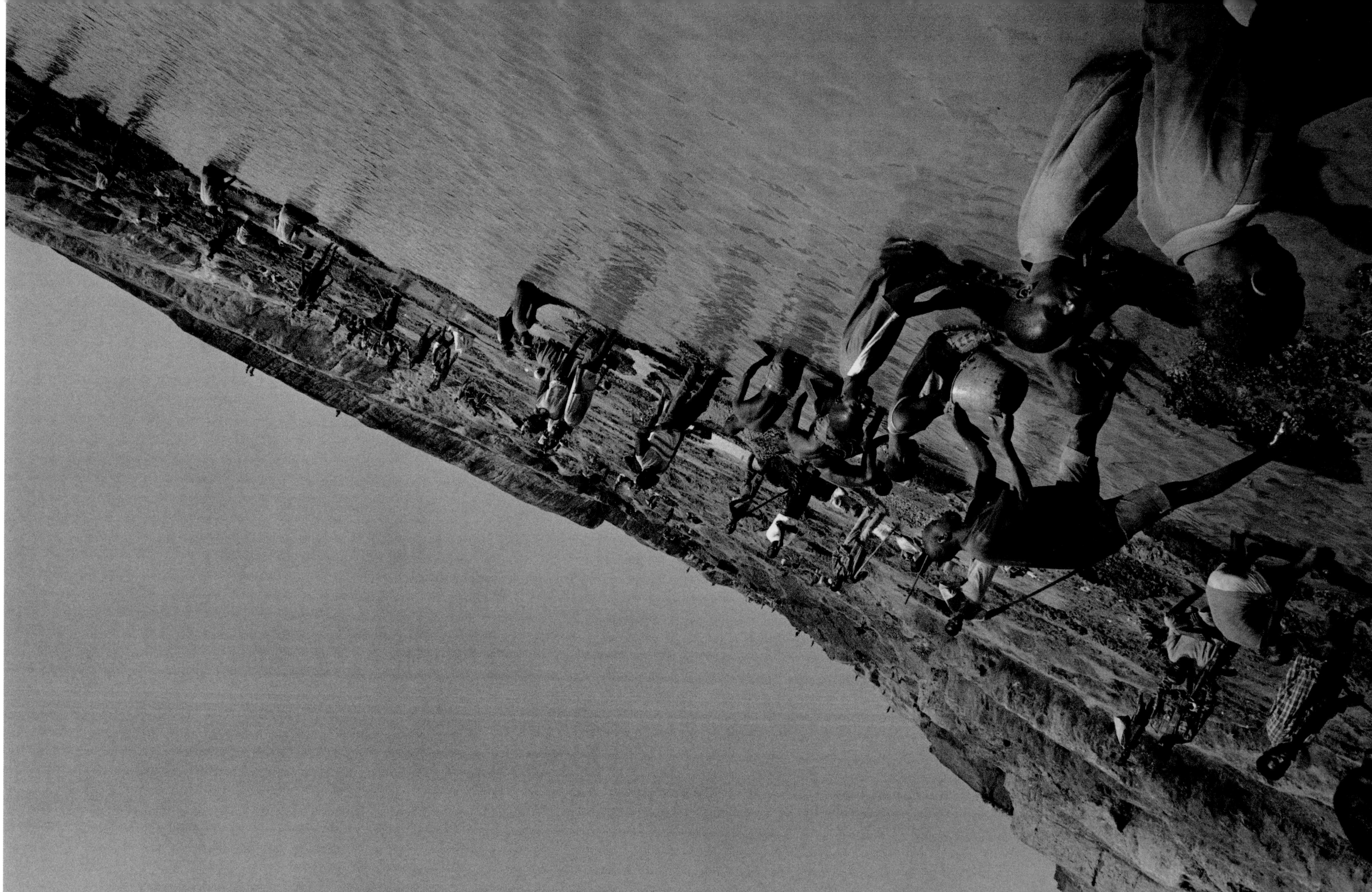

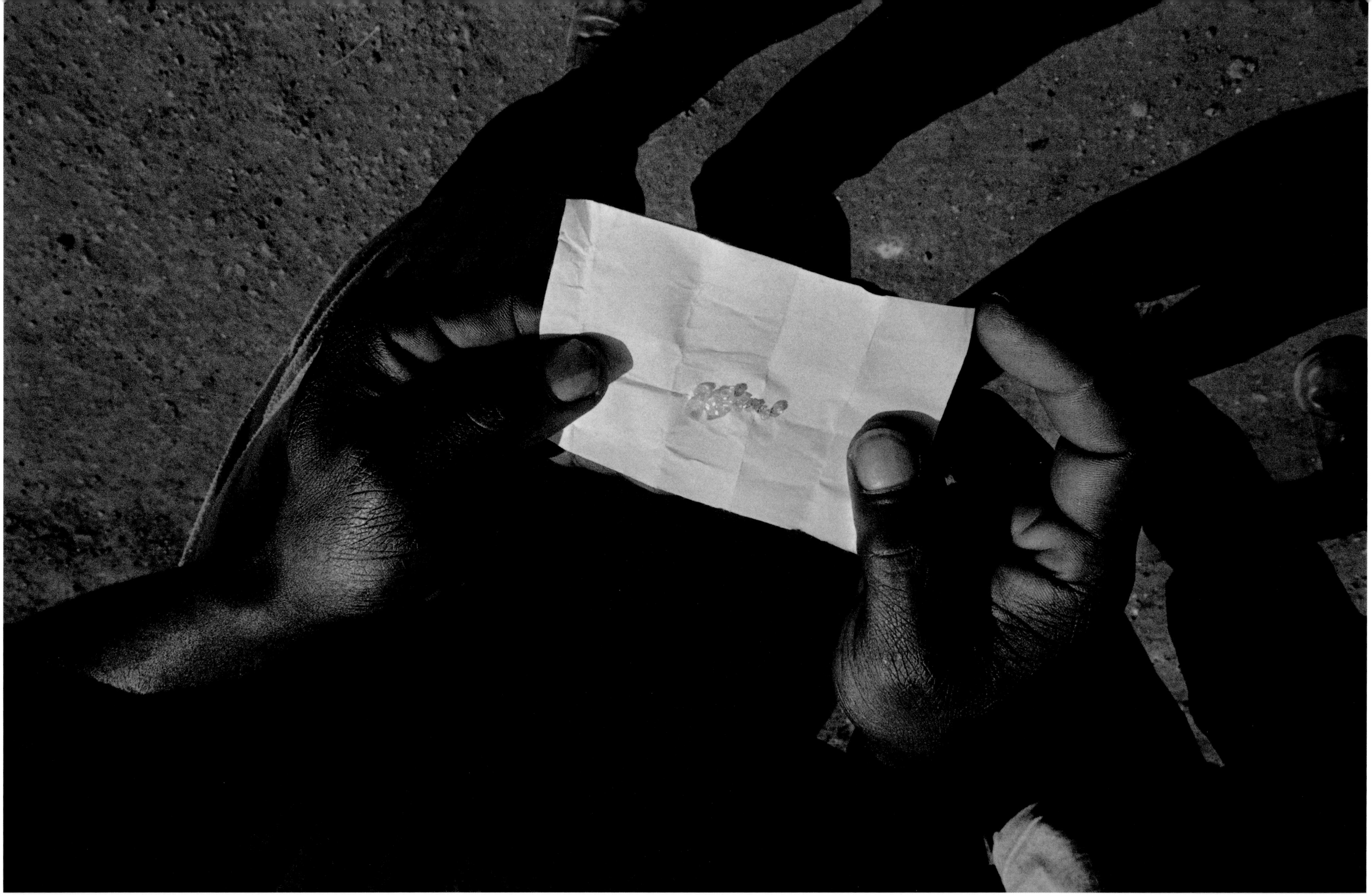

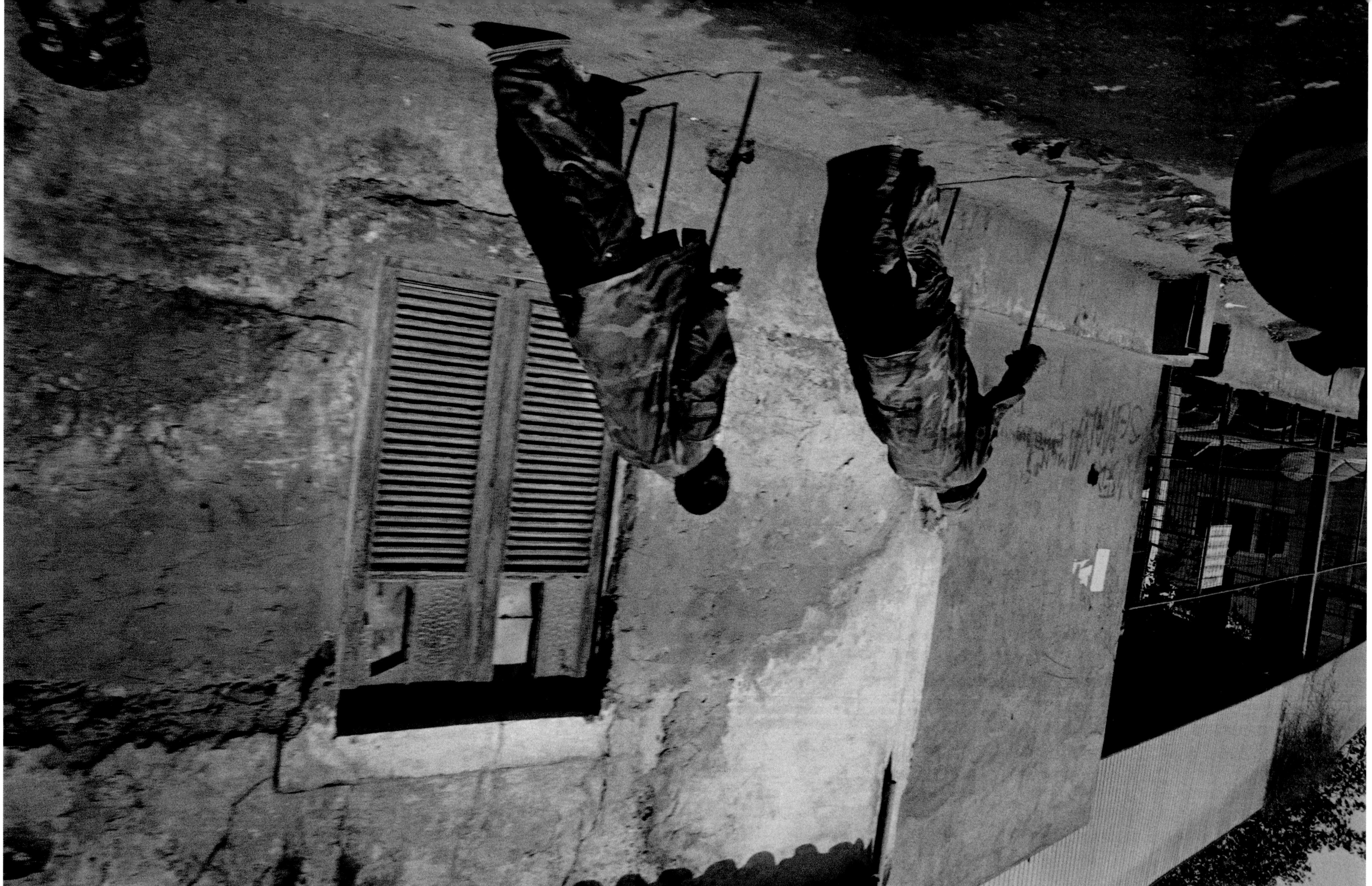

_ 171 WOUNDED BY MINES, FORMER GOVERNMENT SOLDIERS . LUANDA

* MARE HYRCANIUM, OR KHAZAR SEA.

* LARGEST BODY OF INLAND WATER IN THE WORLD.
* **SIZE:** 144,000 sq mi (373,000 sq km).
* 700 MILES LONG.
* **MAXIMUM DEPTH:** 3,200ft (980m).
* CASPIAN SURFACE LIES 92 FT (28M) BELOW SEA LEVEL.

* **BORDERS:** Kazakhstan (NE), Turkmenistan (SE), Iran (S), Azerbaijan (SW) and Russia (NW).

* THE EYES OF THE WESTERN OIL DRIVEN ECONOMIES LOOK UPON THIS BODY OF WATER.

Home to some of the largest.hydrocarbon reserves in the world yet to be tapped.
Contains a proven supply of oil measured between 10-32 billion barrels.
Contains a possible supply of oil estimated at 233 billion barrels.

* BY 2040 ITS OIL DEPOSITS WILL BE ONE OF THE LAST REMAINING ON EARTH.

* "OIL FEEDS THE ARTERY OF THE LIFE OF THE CRUSADER NATION." — *Al Qaeda*

XII :

the caspian sea _ in the face of unsustainable exploitation

_ *Section*

XII

IN THE FACE OF
UNSUSTAINABLE EXPLOITATION

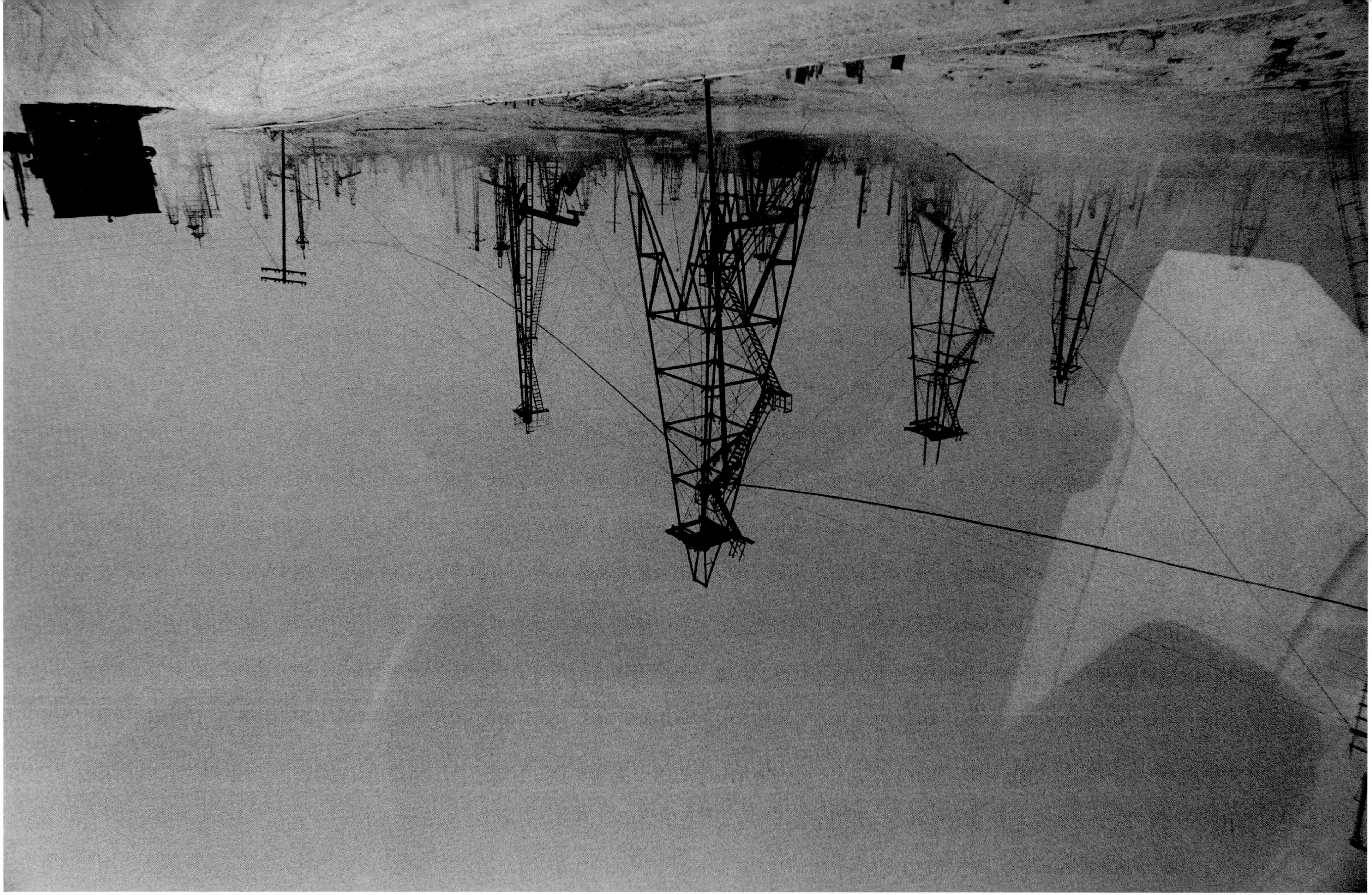

_ 175 OIL FIELD . BAKU

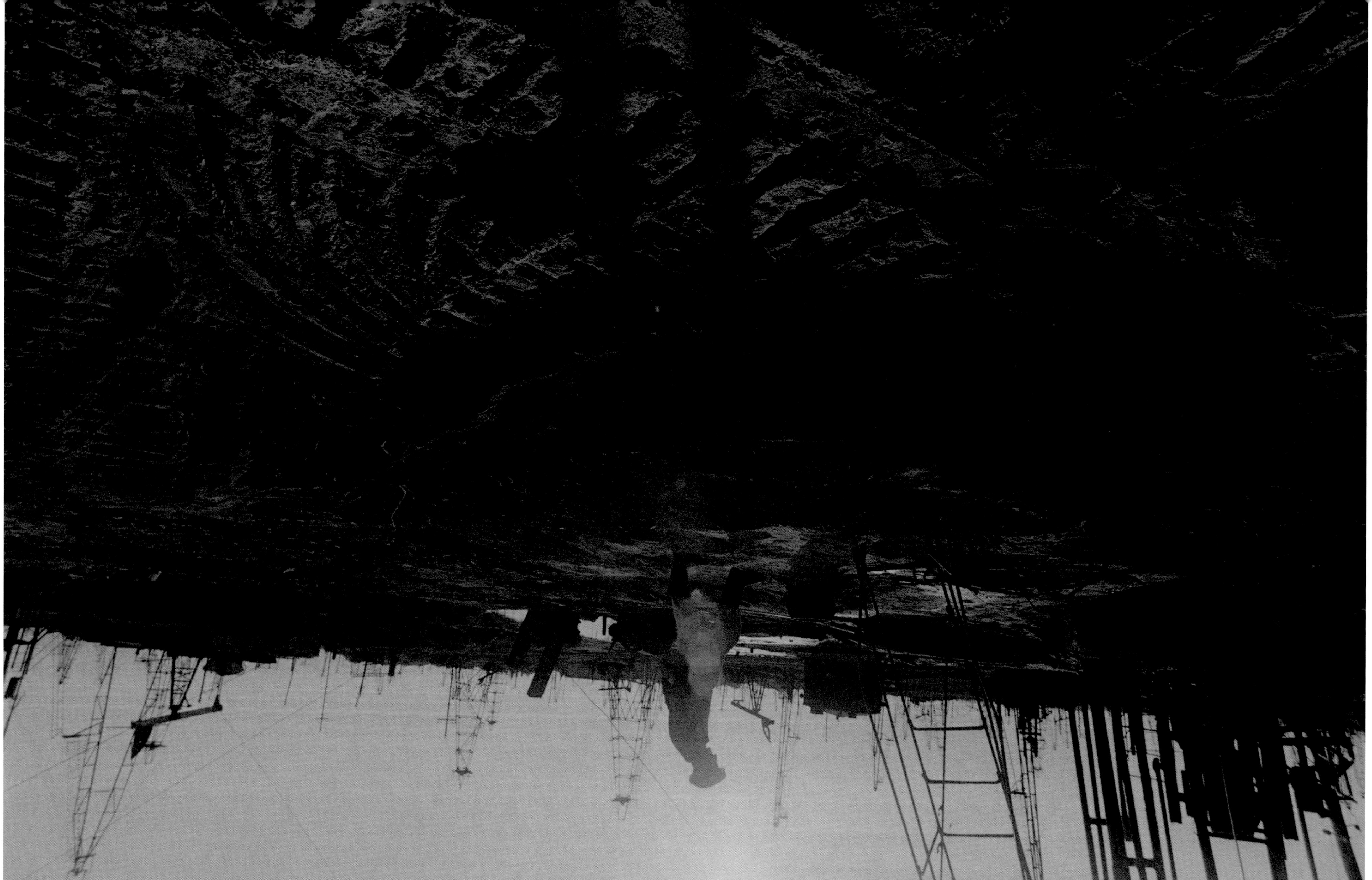

_ 177 OIL FIELD WORKER . BAKU
_ 178 SOVIET MONUMENT . SEASIDE
_ 179 POLLUTED CASPIAN SEA
_ 180 OIL PUMPING STATION . CASPIAN SEA
_ 181 OIL WORKER, OIL ROCKS . CASPIAN SEA

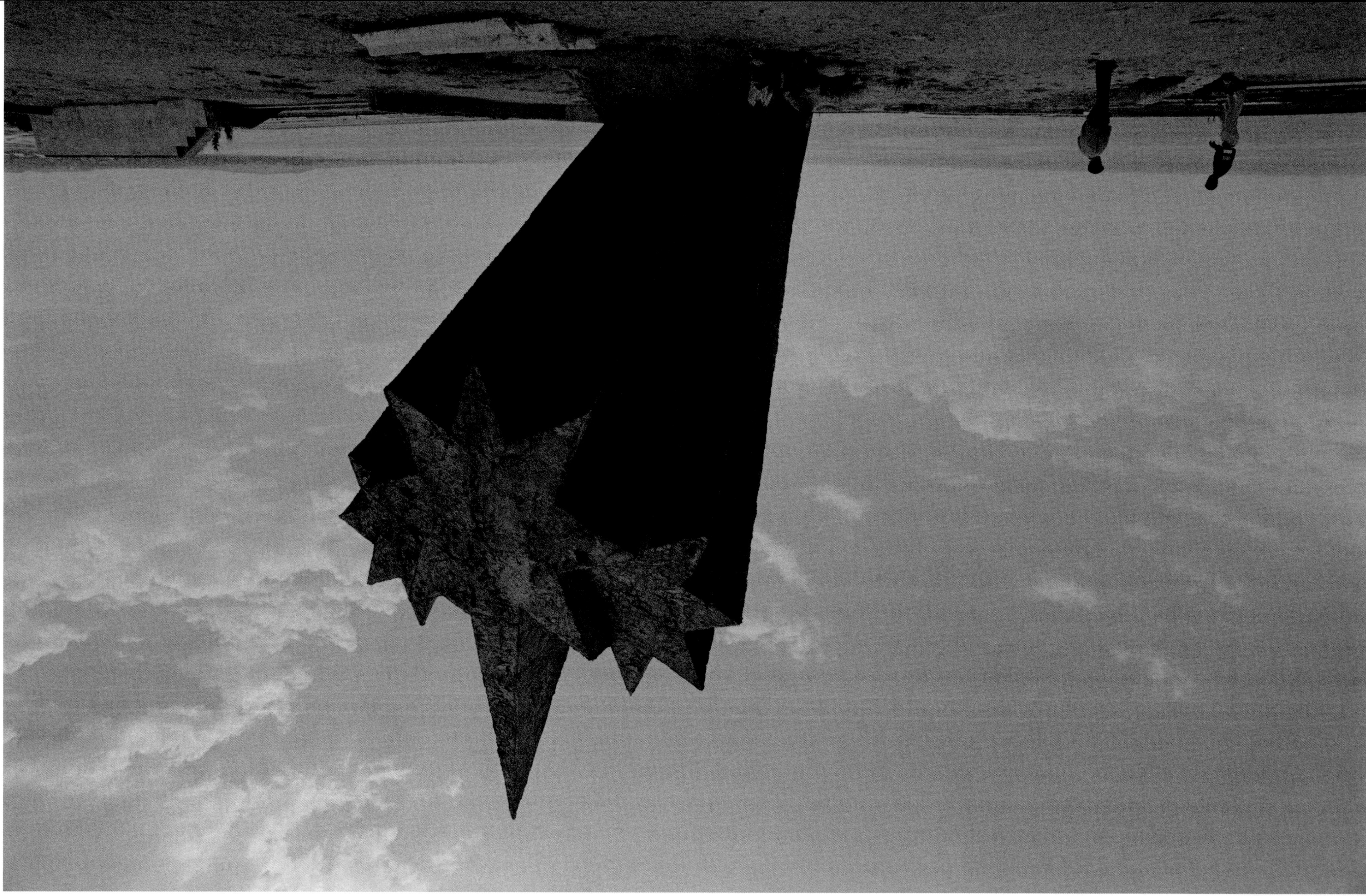

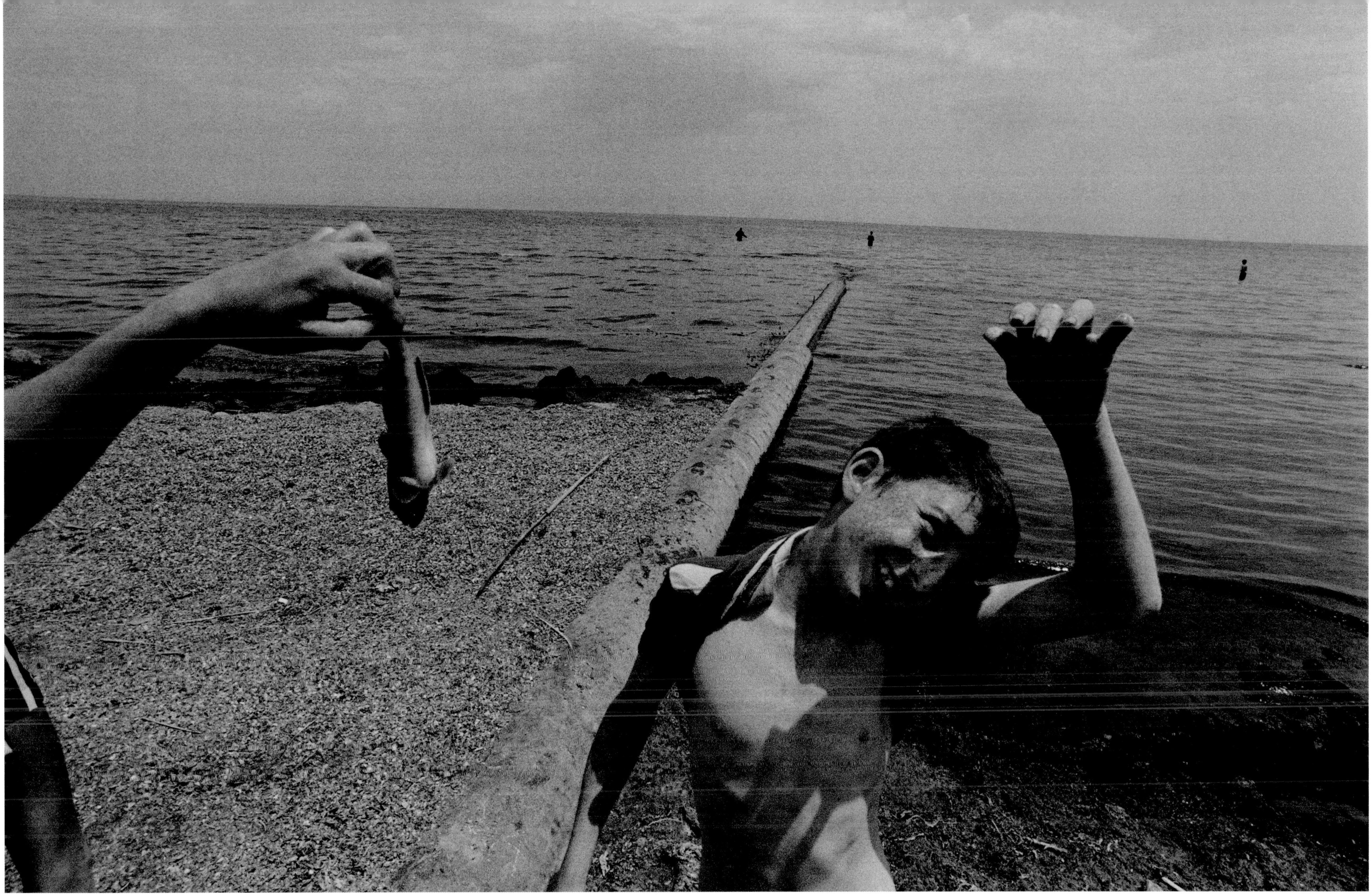

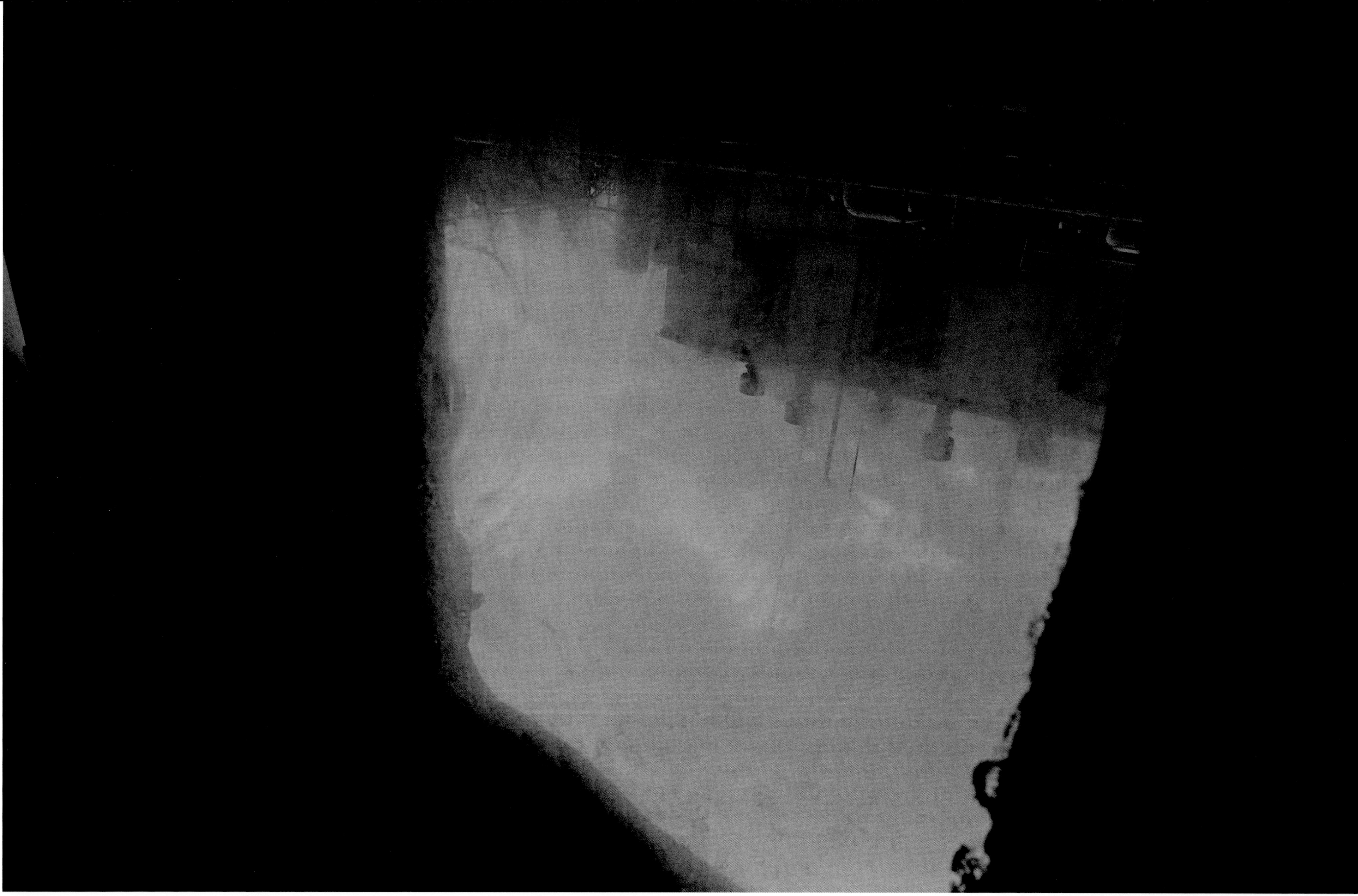

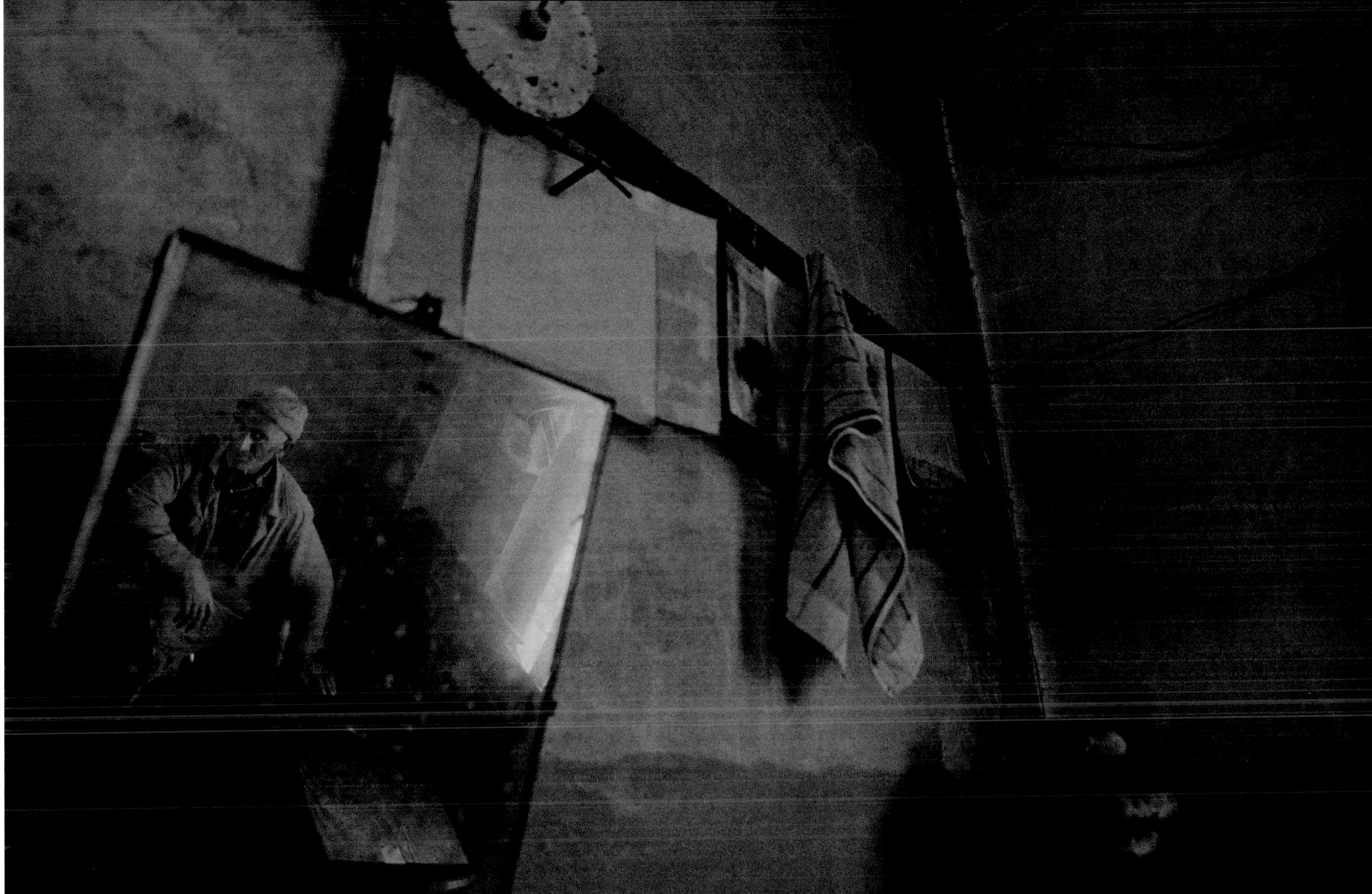

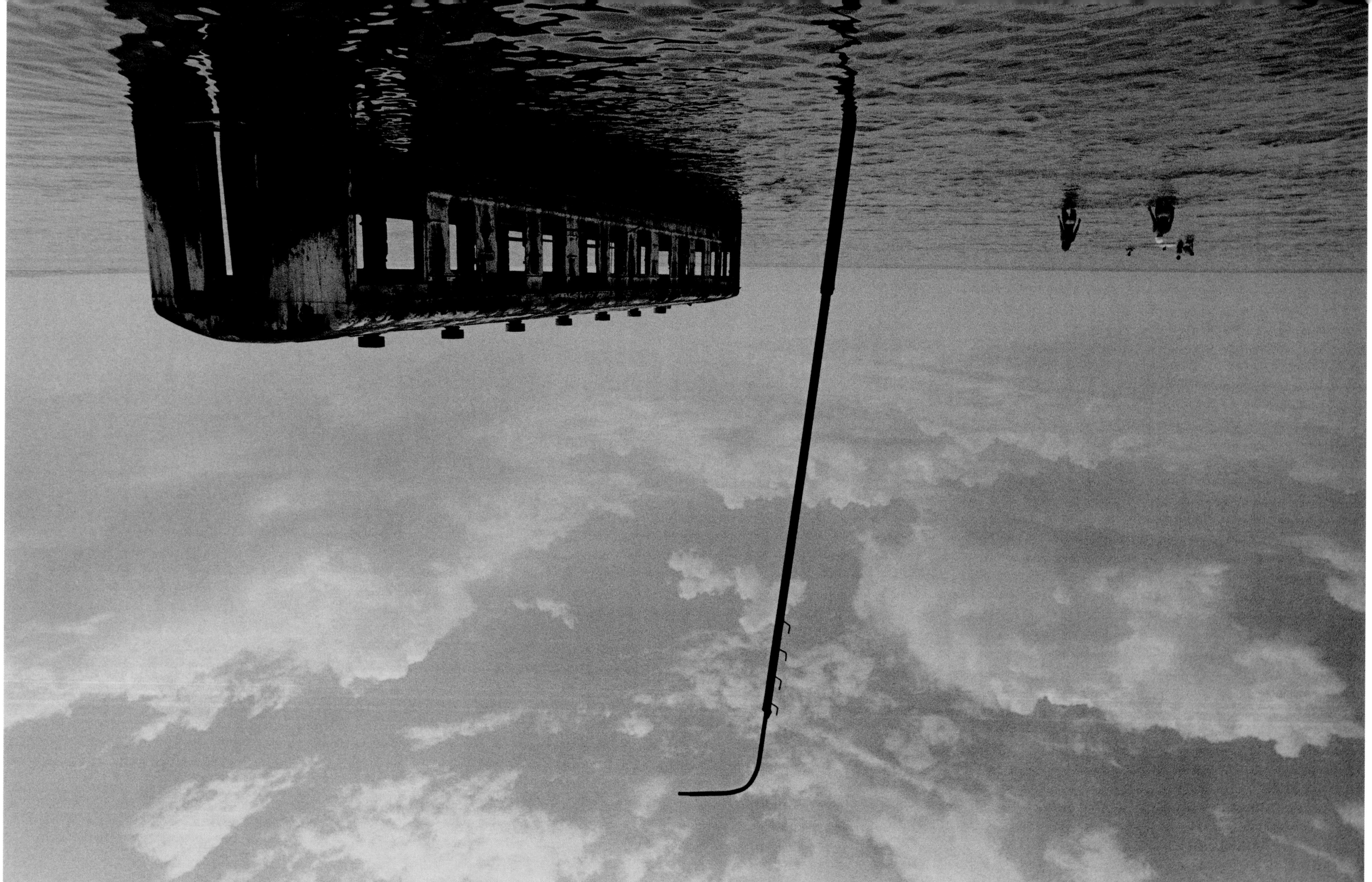

_ 183 OIL LYING ON THE SURFACE OF THE SEA PREVENTS EVAPORATION AND SO FLOODS THE LAND

TO THE CONNOISSEUR IT IS [BLACK GOLD] •

One hundred and seventy-five dollars per ounce or 5 teaspoons full.
The demand for Beluga caviar is described as infinitely **INELASTIC.**

No matter the price,
Beluga will be served.

Inelastic demand guarantees supply.
And inelastic demand will suck the Caspian Sea dry of its most prestigious commodity.

Beluga is not a luxury;
it's status —
those with
the economic power
will eat
black gold
until there is
none left.

Nothing compares to Beluga caviar — nothing can take its place.

XIII :

azerbaijan _ black market, black gold

_ *Section*

XIII

BLACK MARKET, BLACK GOLD

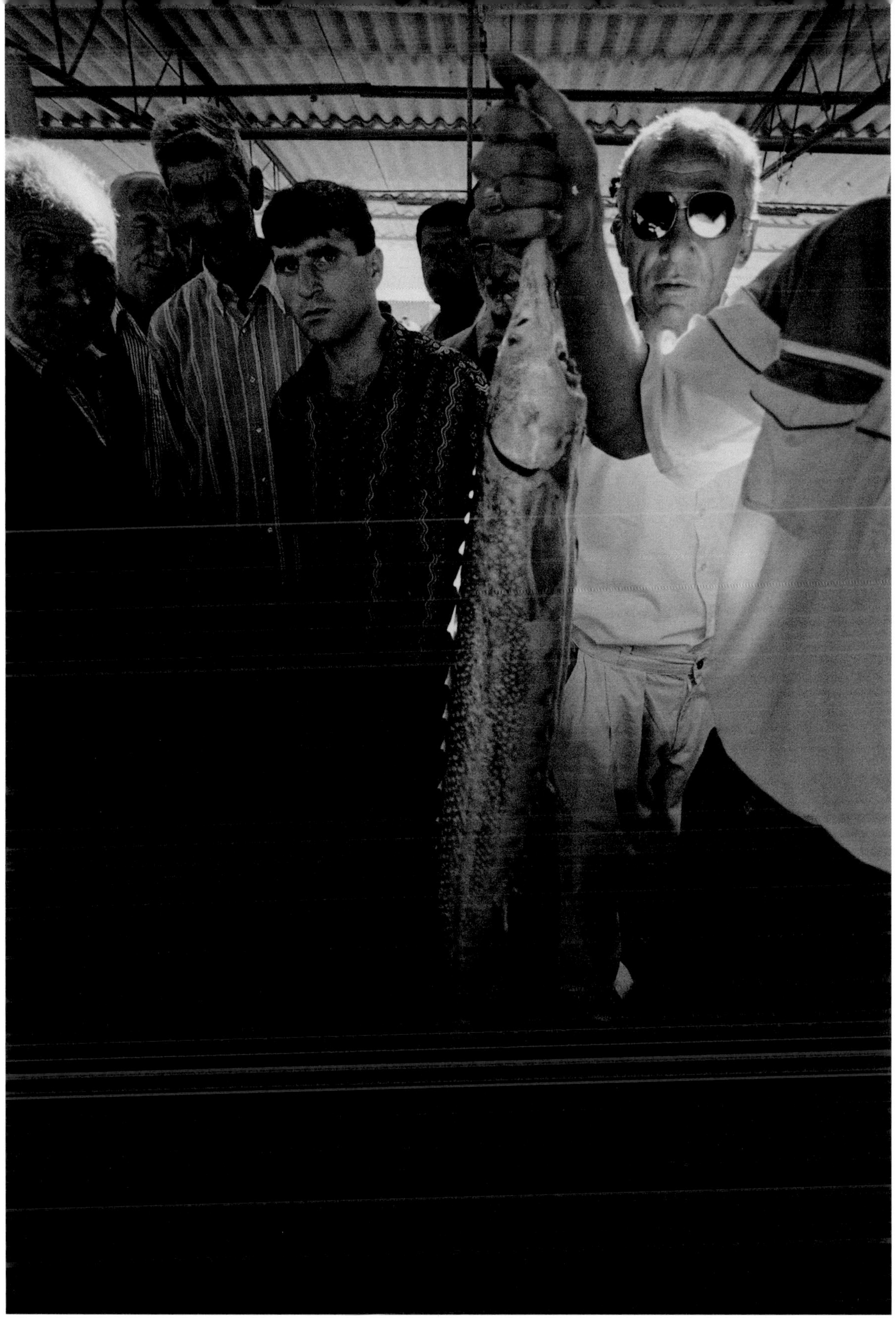

_ 187 CAVIAR POACHERS HOLD A STURGEON AT A BLACK MARKET

Thirty years of civil war has
reduced Cambodia's
tropical rainforests,
once covering 70% of the country's landmass,
to an estimated 30%.

BETWEEN 1975 AND 1986, POL POT'S KHMER ROUGE CUT DOWN BETWEEN 10-20 MILLION DOLLARS A MONTH IN ILLEGAL TIMBER.

In 1975, one AK-47 in Cambodia's Phnom Penh cost 1,000 Thai baht ($25) ... one bullet 5 baht (10¢).

In 1975, **POL POT** began his purification of Cambodian society.

[YEAR ZERO] WAS TO BE THE BEGINNING OF HIS UTOPIAN DREAM .

That year an estimated 2 million people were liquidated through ill-treatment,
over-work and execution by 25 dollar guns and dime-a pop bullets.

MOST VICTIMS WERE DISPOSED OF IN HUGE EXPANSES OF ARABLE LAND. THESE BECAME KNOWN AS THE [KILLING fiELDS] .

Between 1975 and 1986 the weapons
that were bought through the process of illegal logging
helped bring about a genocide
that wiped away one quarter of the country's population.

From 1986 to
present day
the ecological
massacre
goes on.

CAMBODIAN WOOD IS USED IN THE MANUFACTURE OF PATIO FURNITURE.

A Waikiki sun lounger made of Balau wood costs £49.95 ($90.00).
Nice on the patio,
lovely in the garden,
perfect for
the killing fields.

cambodia _ cut in cambodia, made in vietnam, sold in the west

_ *Section*

XIV

CUT IN CAMBODIA, MADE IN VIETNAM, SOLD IN THE WEST

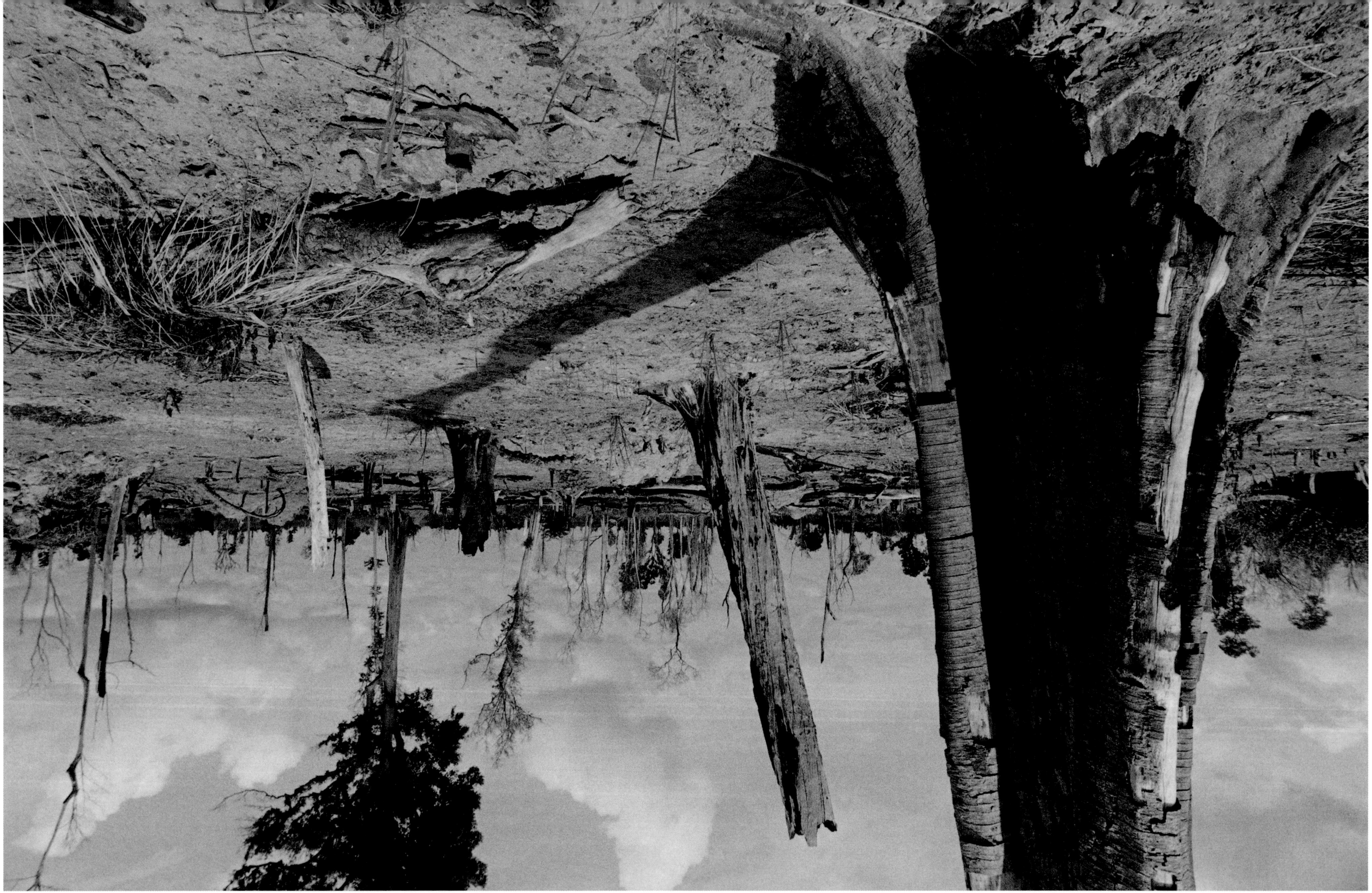

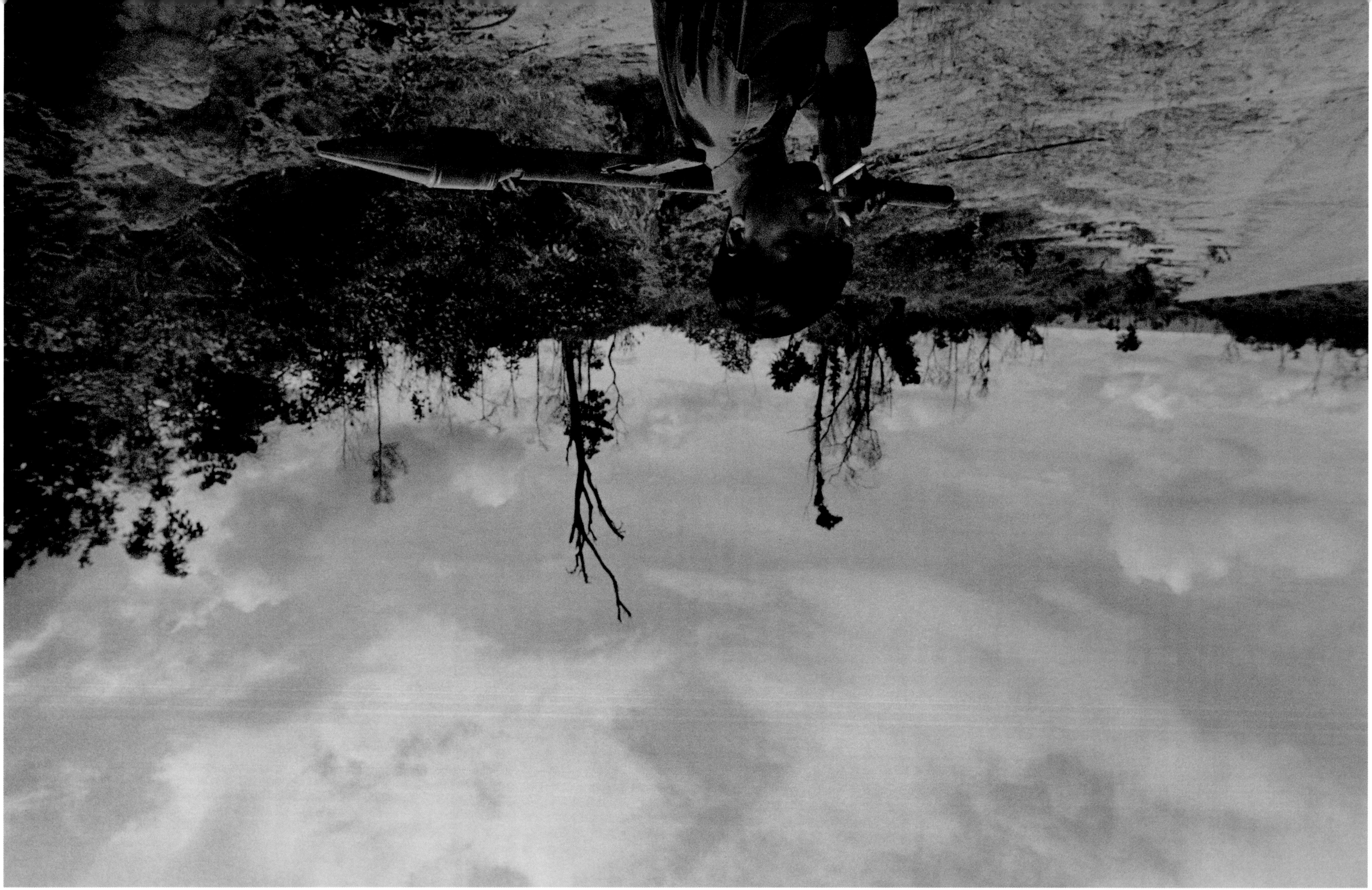

_ 193 FORMER KHMER ROUGE SOLDIER CARRIES CHINESE B40 GRENADE LAUNCHER

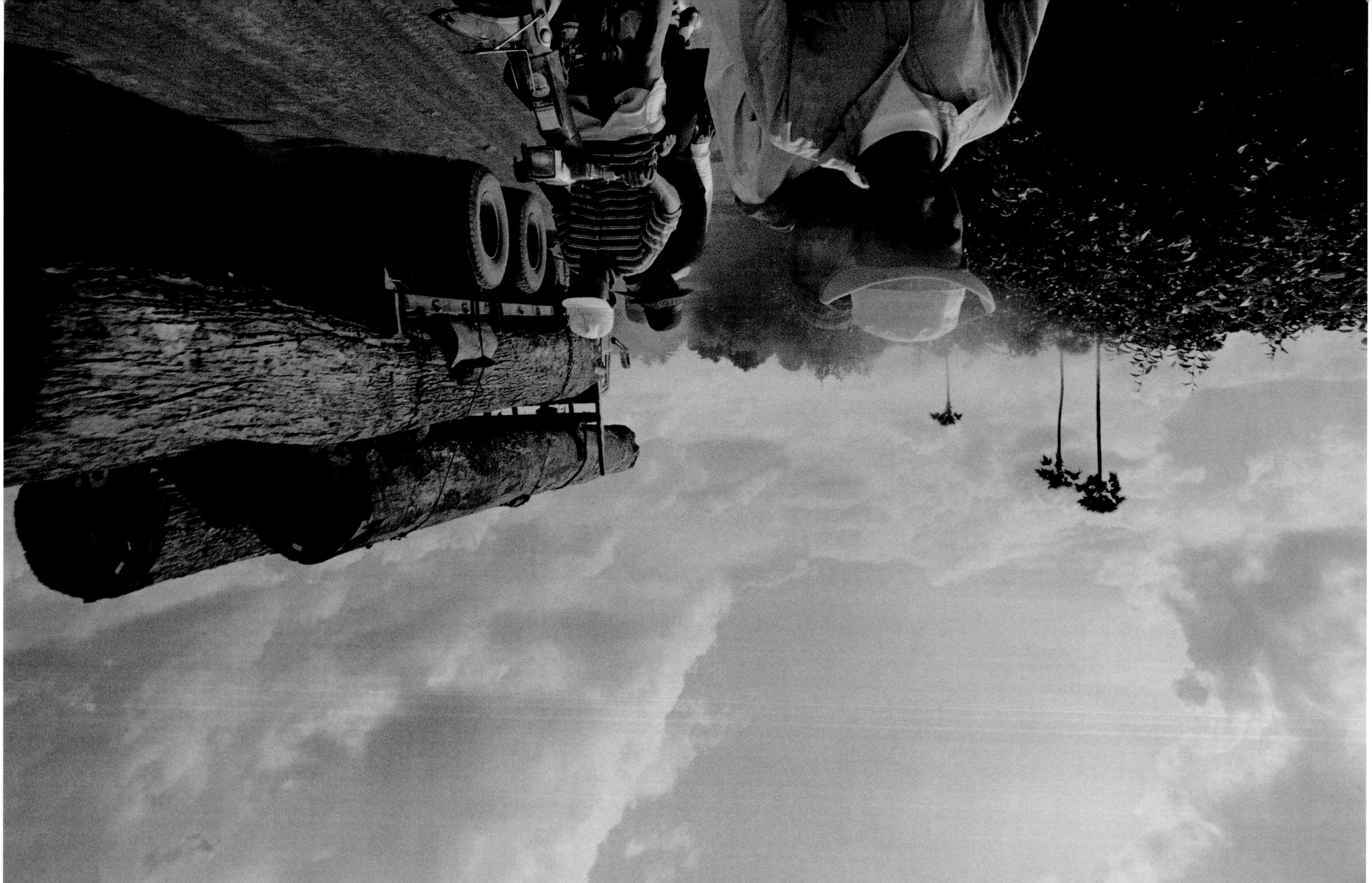

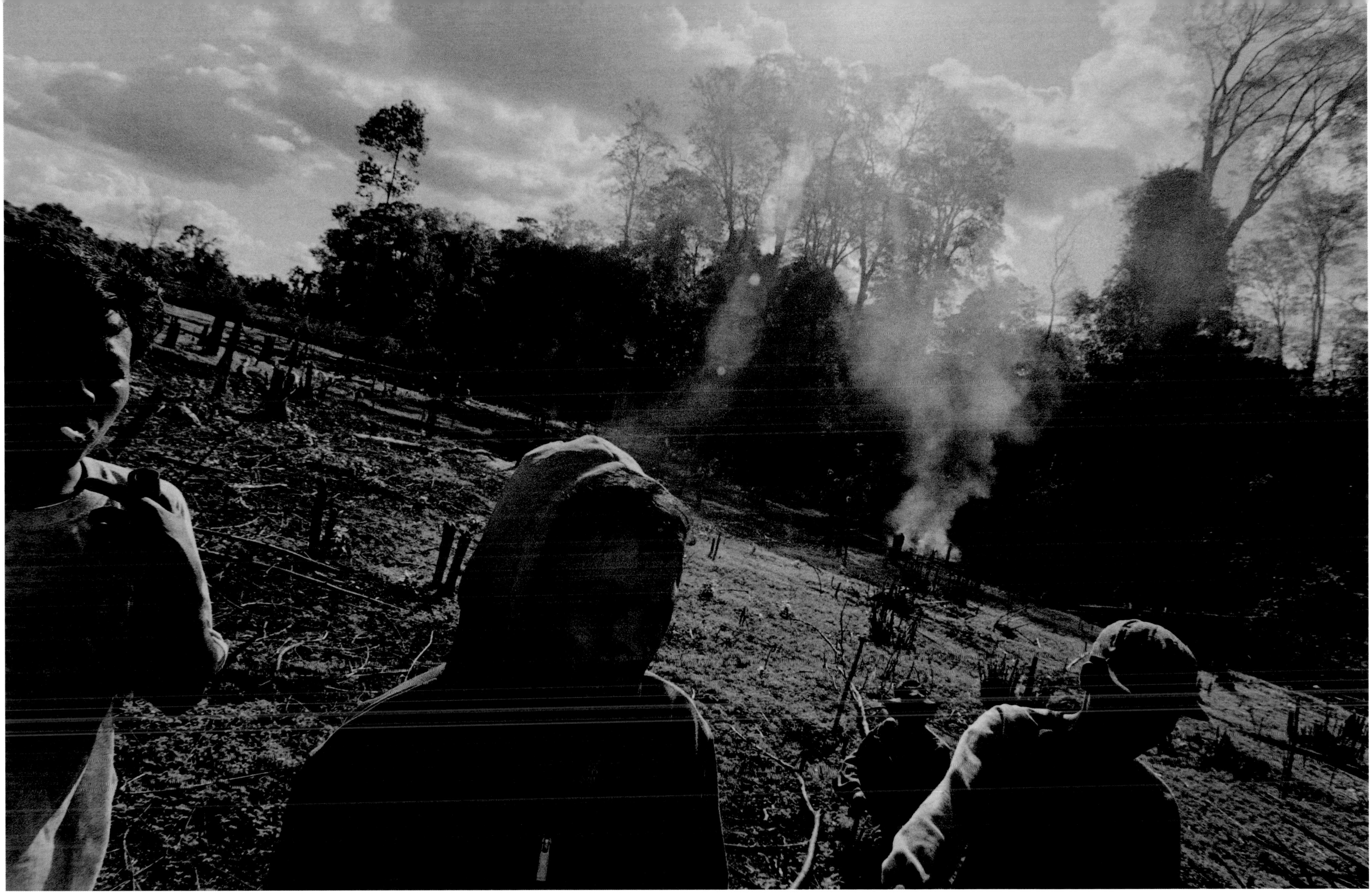

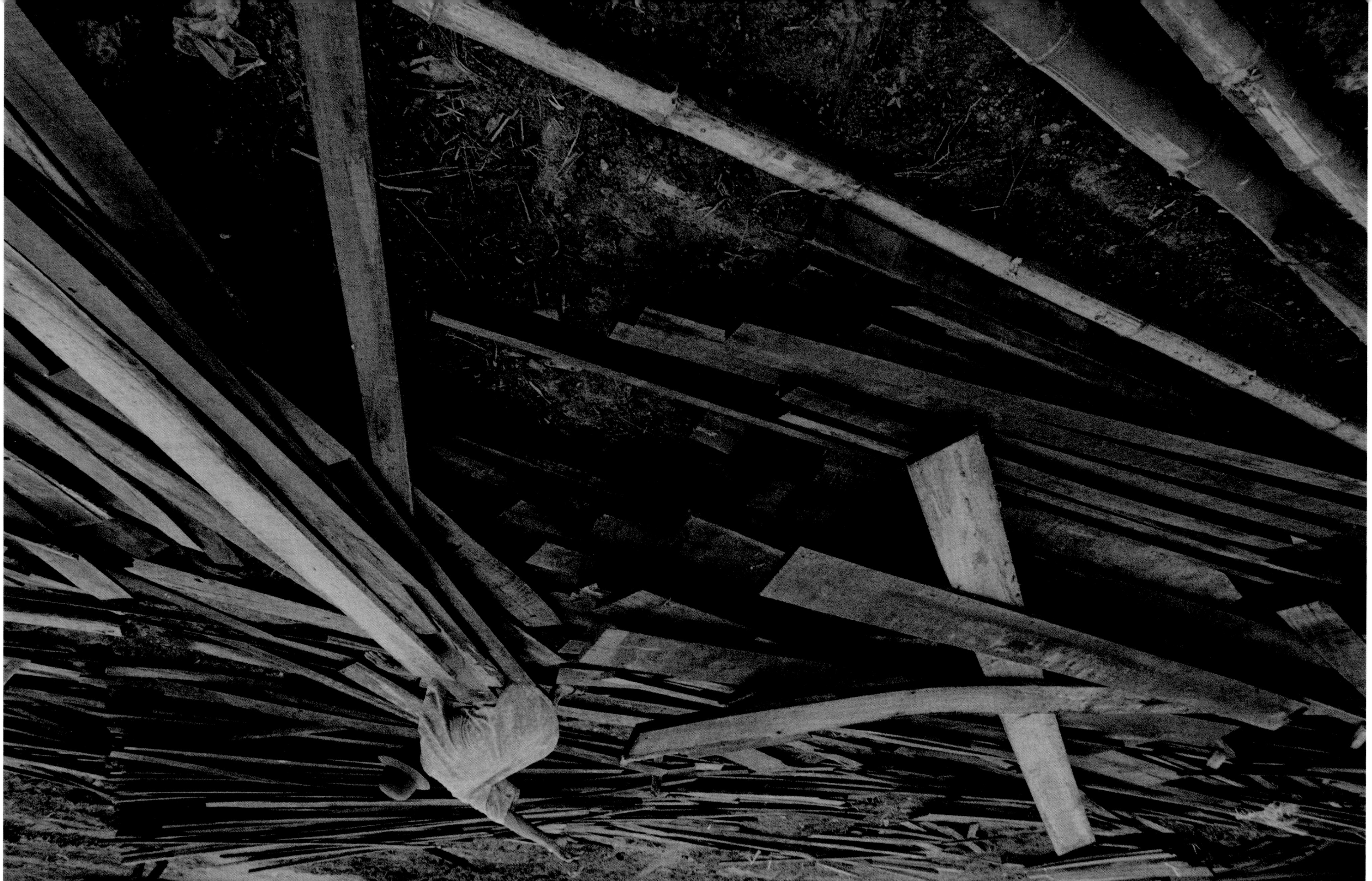

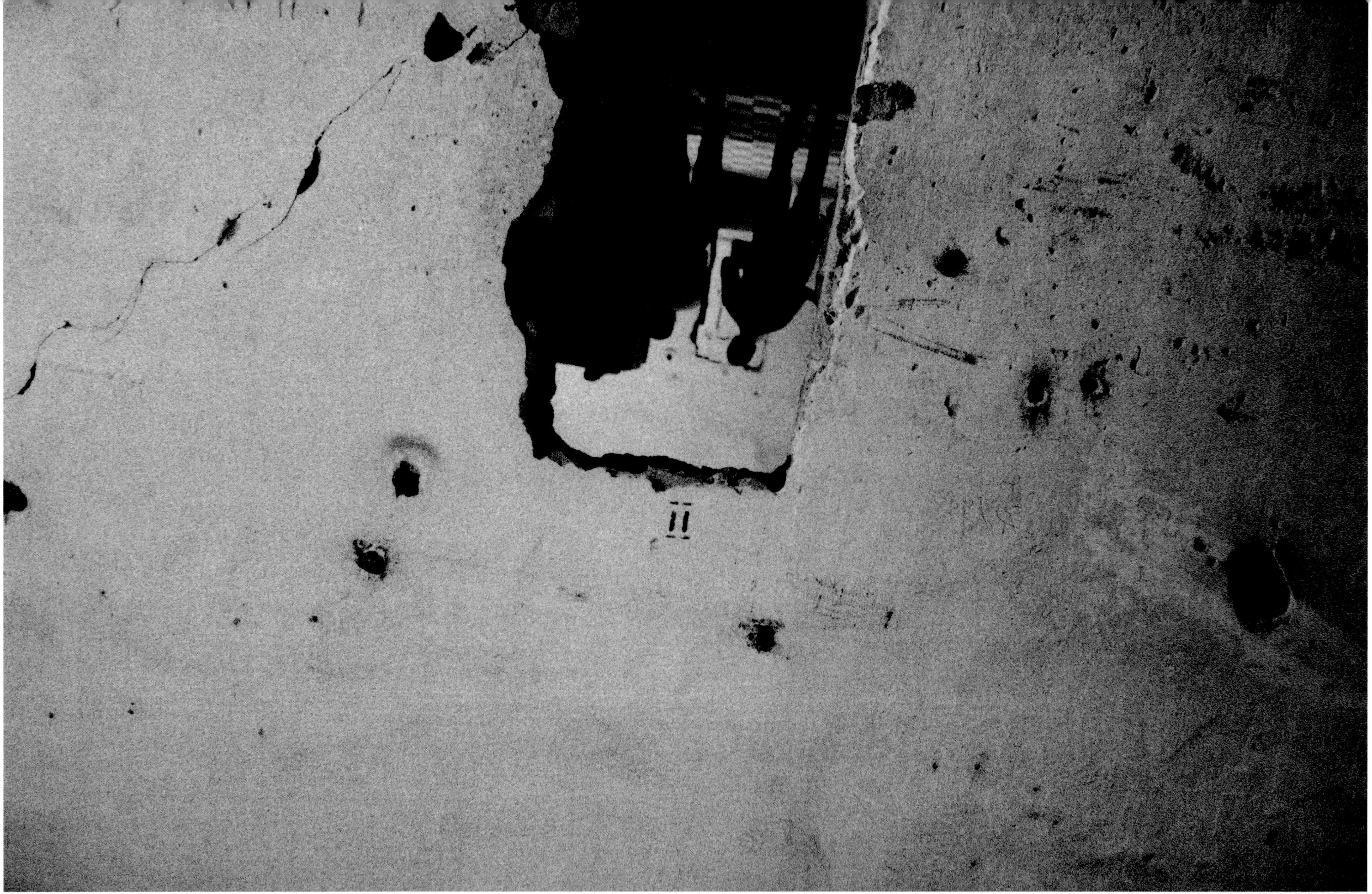
II

_ 201 INFAMOUS S-21 PRISON IN PHNOM PENH
_ 202 PHOTOS OF KHMER ROUGE PRISONER . PHNOM PENH
_ 203 KILLING FIELDS

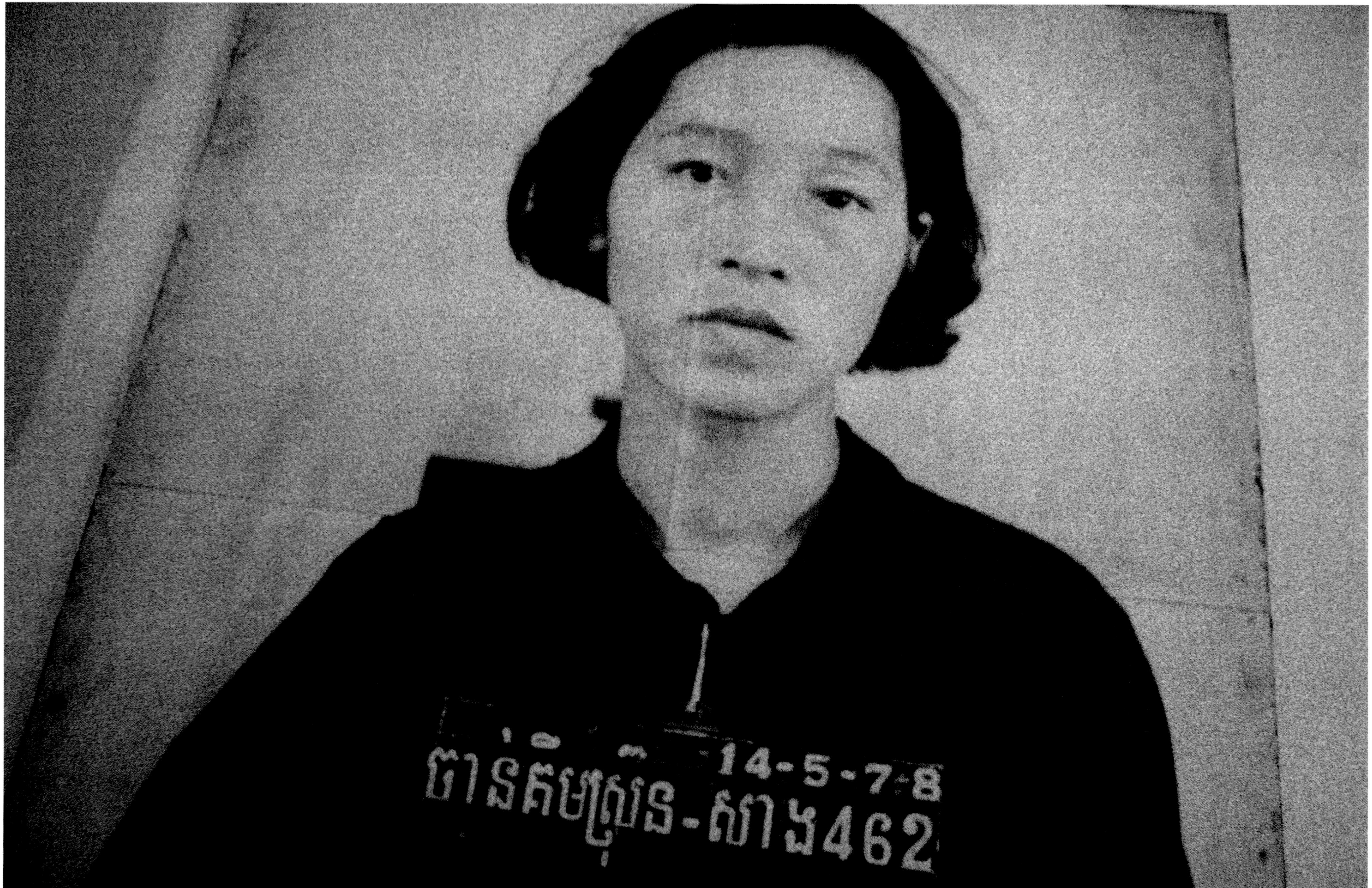
14-5-7-8
ចាន់គឹមស្រ៊ុន-សាង462

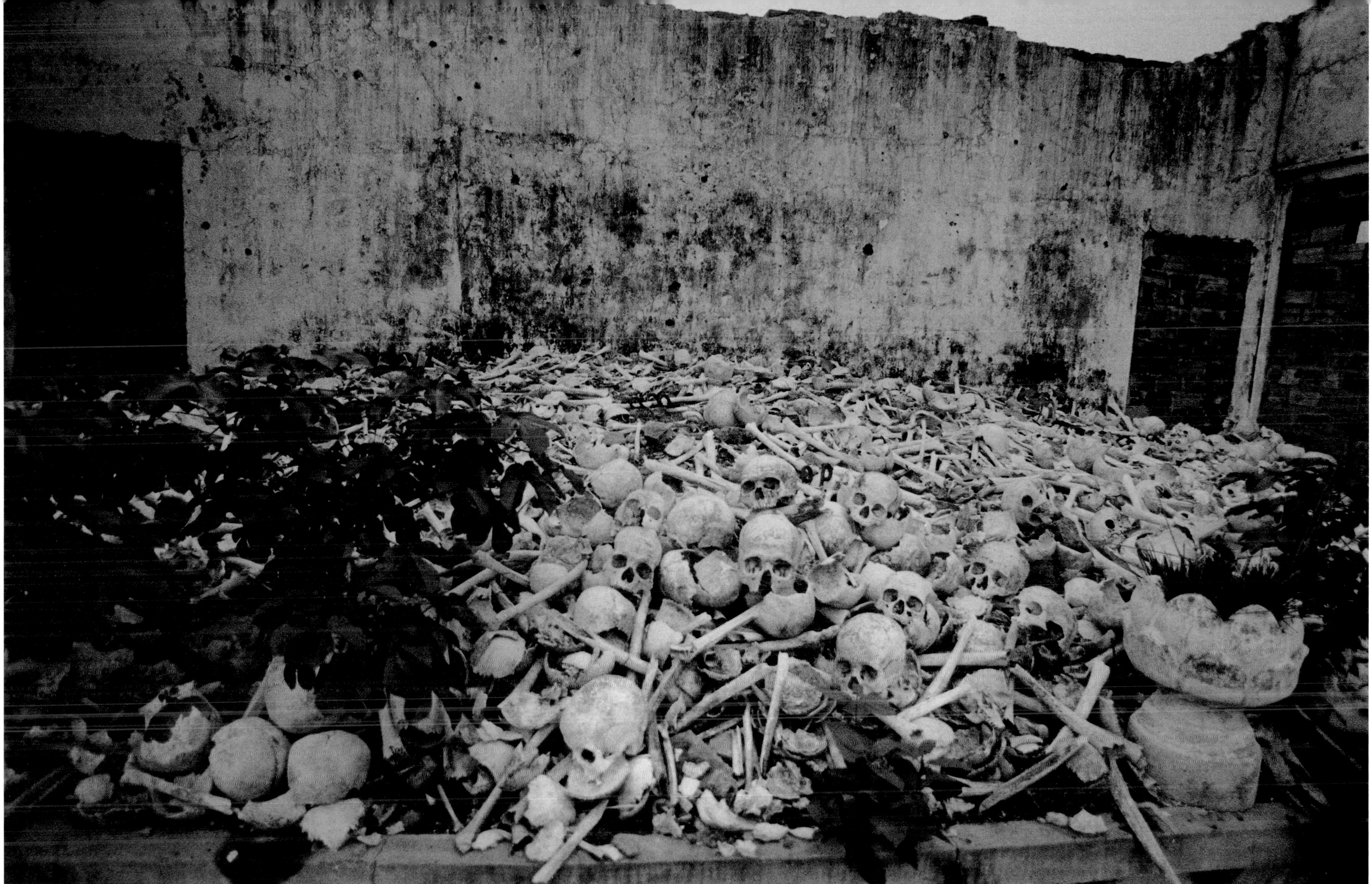

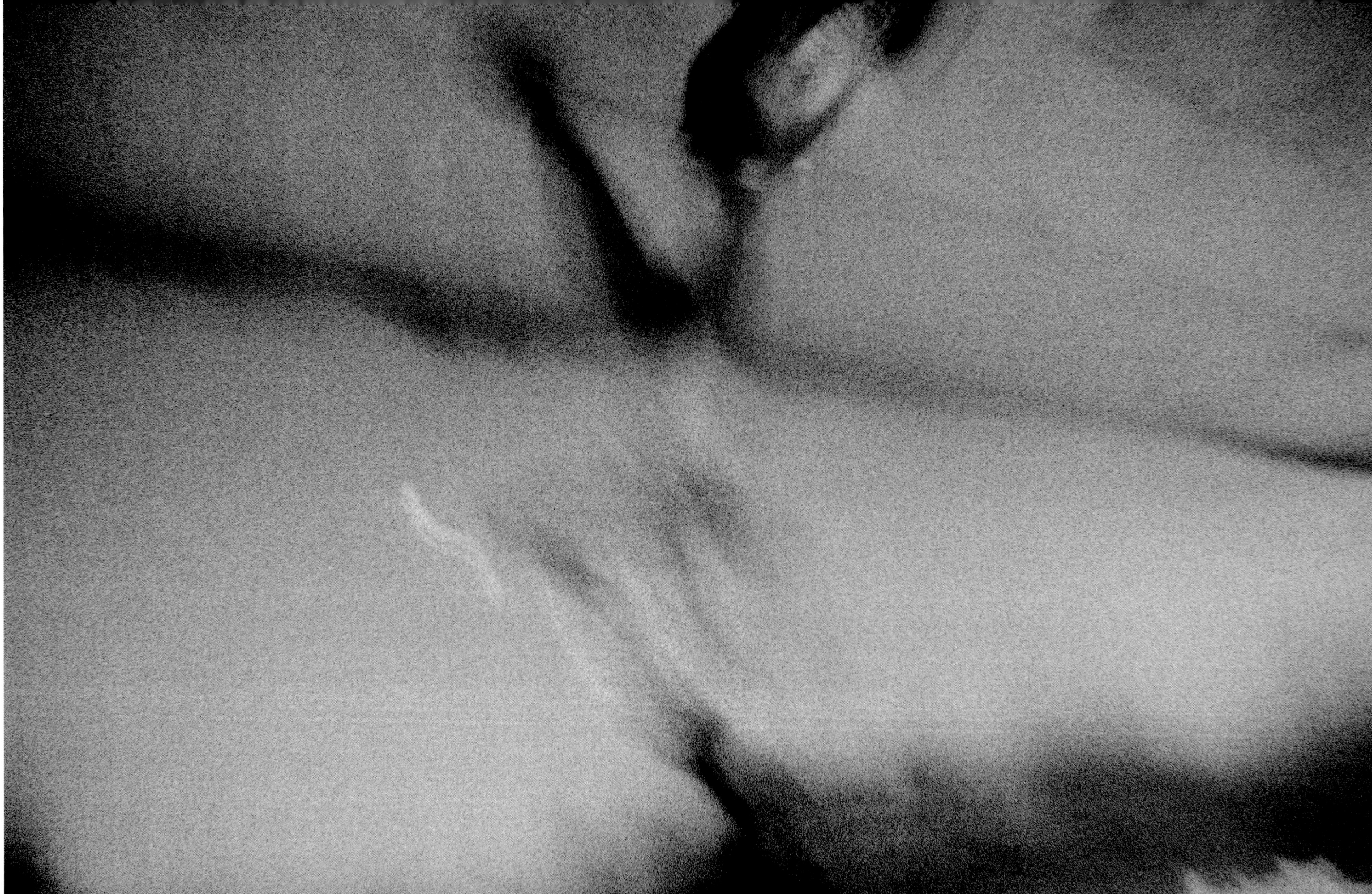

WAR PROSECUTES [THE LAND] , AND [THE LAND] YIELDS BUT FOR A MOMENT .

Faces peer through shards of window in buildings battered and bombed. Liberated people can't quite see what freedom. really looks like,
but stars and stripes and union jacks all paint a pretty picture.

Through SHOCK AND AWE, blast and quake, a culture is bled then swallowed whole.
The ground has ceased to rumble from the laser-guided storm.

The Nemesis has dealt its blow in its theatre of operations.
The second act will soon commence once the dead are dragged off center stage.

WAR PROSECUTES [THE LAND] , AND [THE LAND] YIELDS IN DEFENSE OF THOSE IT KEEPS .

Trenches deep with burning oil send signals: "stay away!" The fuel-injected infantries won't stop for pity sake.
Their orders are to come, come and keep on coming until the dictator's heavy hand has lost his iron grip.

Justice is what the invaders believe.
All men are created equal.
Thou shalt not kill they say in church.
And IN GOD WE TRUST their money reads.
And still they come
for those are their orders: come,
come and keep on coming.

WAR PROSECUTES [THE LAND] , AND [THE LAND] YIELDS TO ITS MUTILATION .

TREAD AND HALF-TRACK LEAVE SCARS WHERE SHEEP GRAZED AND CHILDREN PLAYED.
Under a wall the out-matched defender lays buried.
His arm juts from the rubble; his wrist-watch continues to tick.

Civilians flee in single file away towards infinity.
Women wearing only black billow out their homes. Families torn from neighborhoods hold what their arms can carry.
Their broken hearts and memories cannot bear a thing.

WAR PROSECUTES [THE LAND] , AND [THE LAND] CONTINUES .

Sand will repair the wounds,
the gouges left by bombs.
Sand will absorb the blood from the clash of civilizations.
Telegraph poles and pylons will reach from out of the desert towards His smoked-filled heavens.

And with time all things will clear. When the liberators have left and lost their minds.
When orphans have learned to hate blue-eyes. When politics is again in knots.
When similar crusades for similar reasons have been waged in similar somewheres
and when profit has been made and spent and made again. When this has been done and done to death.

WHEN THERE'S NO ONE LEFT TO KILL.

THERE 'LL BE THE MOST BEAUTIFUL SILENCE THAT 'S NEVER BEFORE BEEN HEARD

iraq _ warscapes

_ Section

XV

WARSCAPES

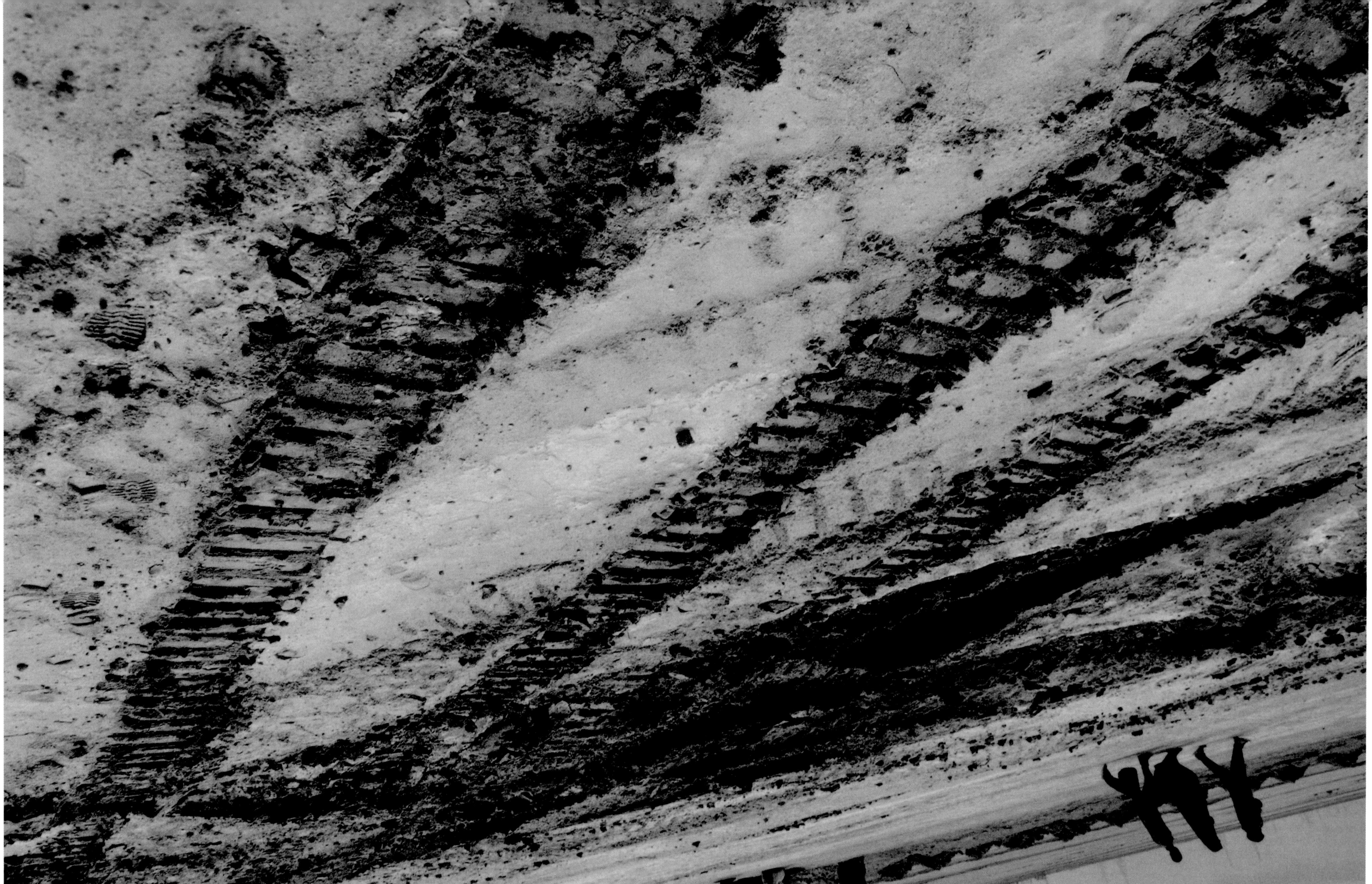

_ 209 TANK TREADS, FLEEING IRAQIS . BASRA
_ 210 MOSAIC OF SADDAM HUSSEIN
_ 211 MOTHER SEARCHES FOR DEAD SON

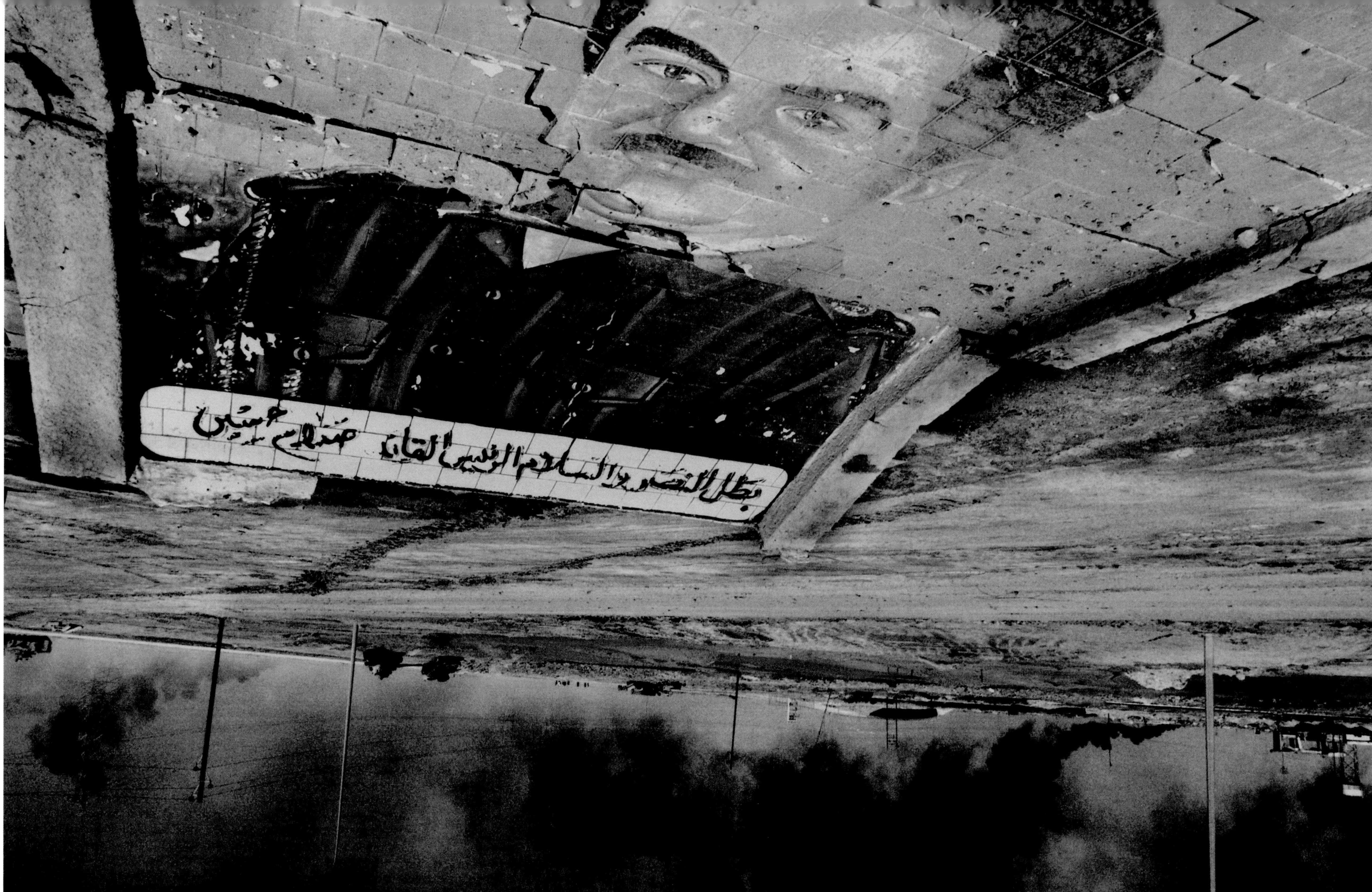
بطل النصر والسلام الرئيس القائد صدام حسين

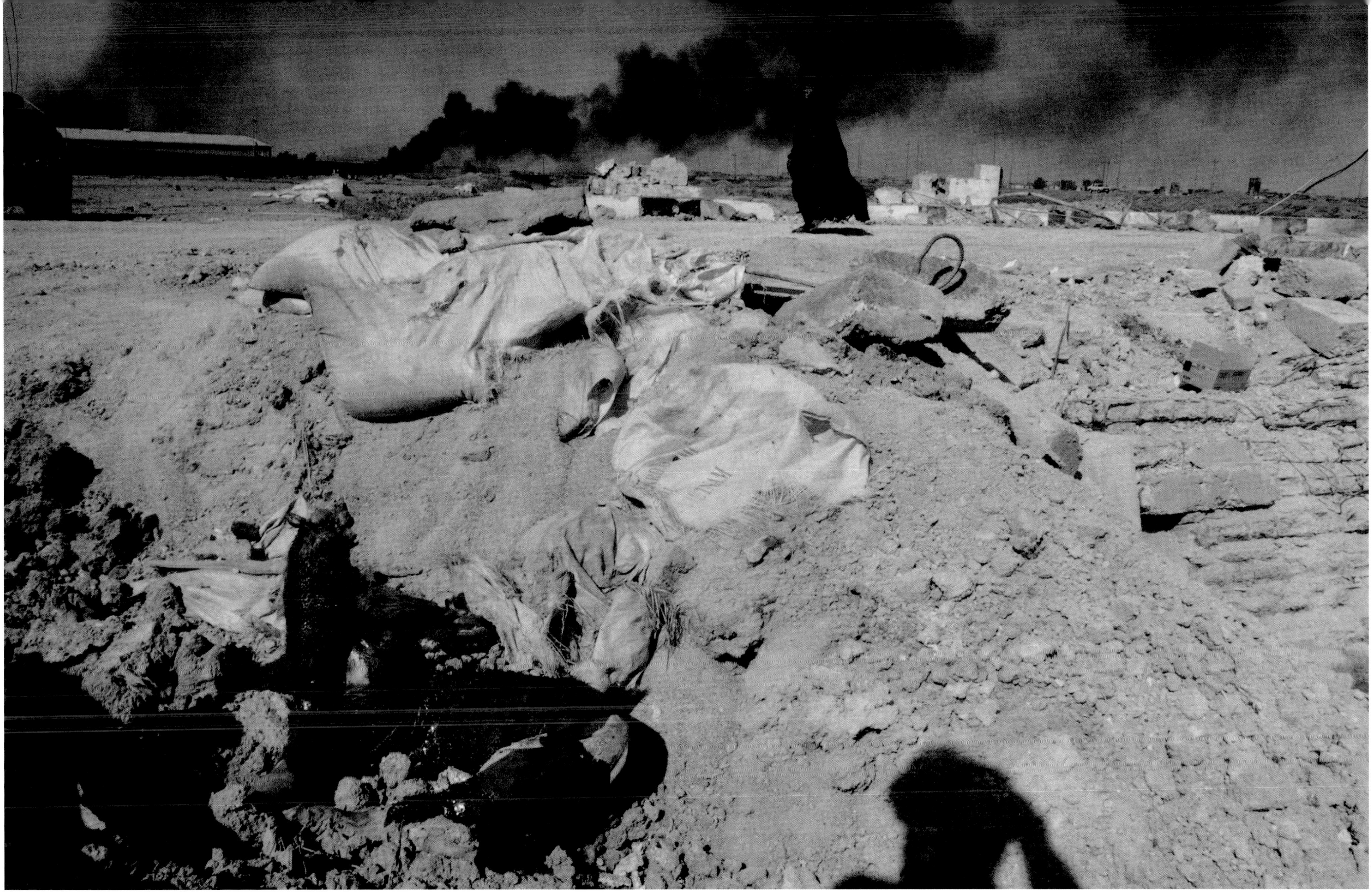

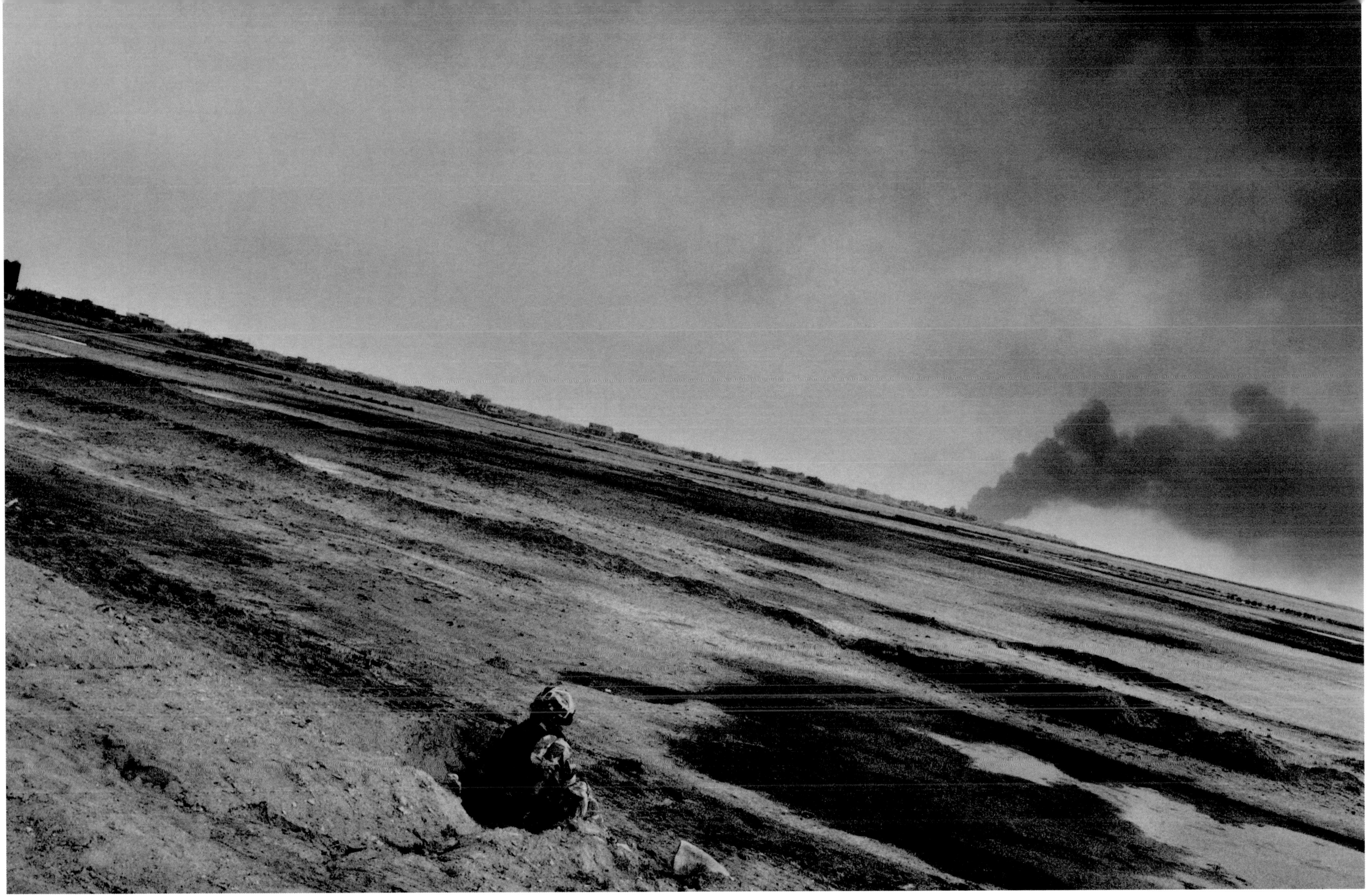

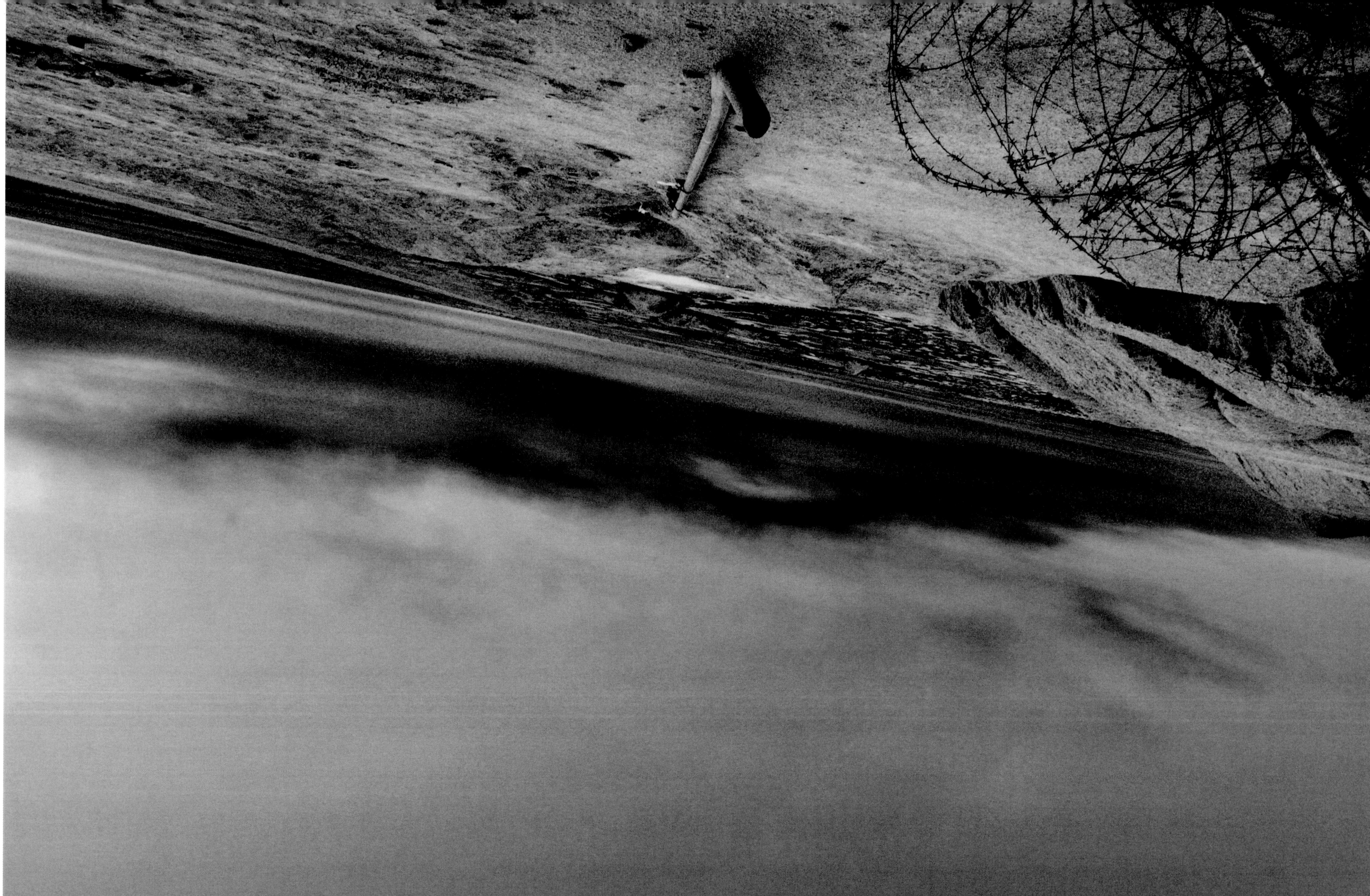

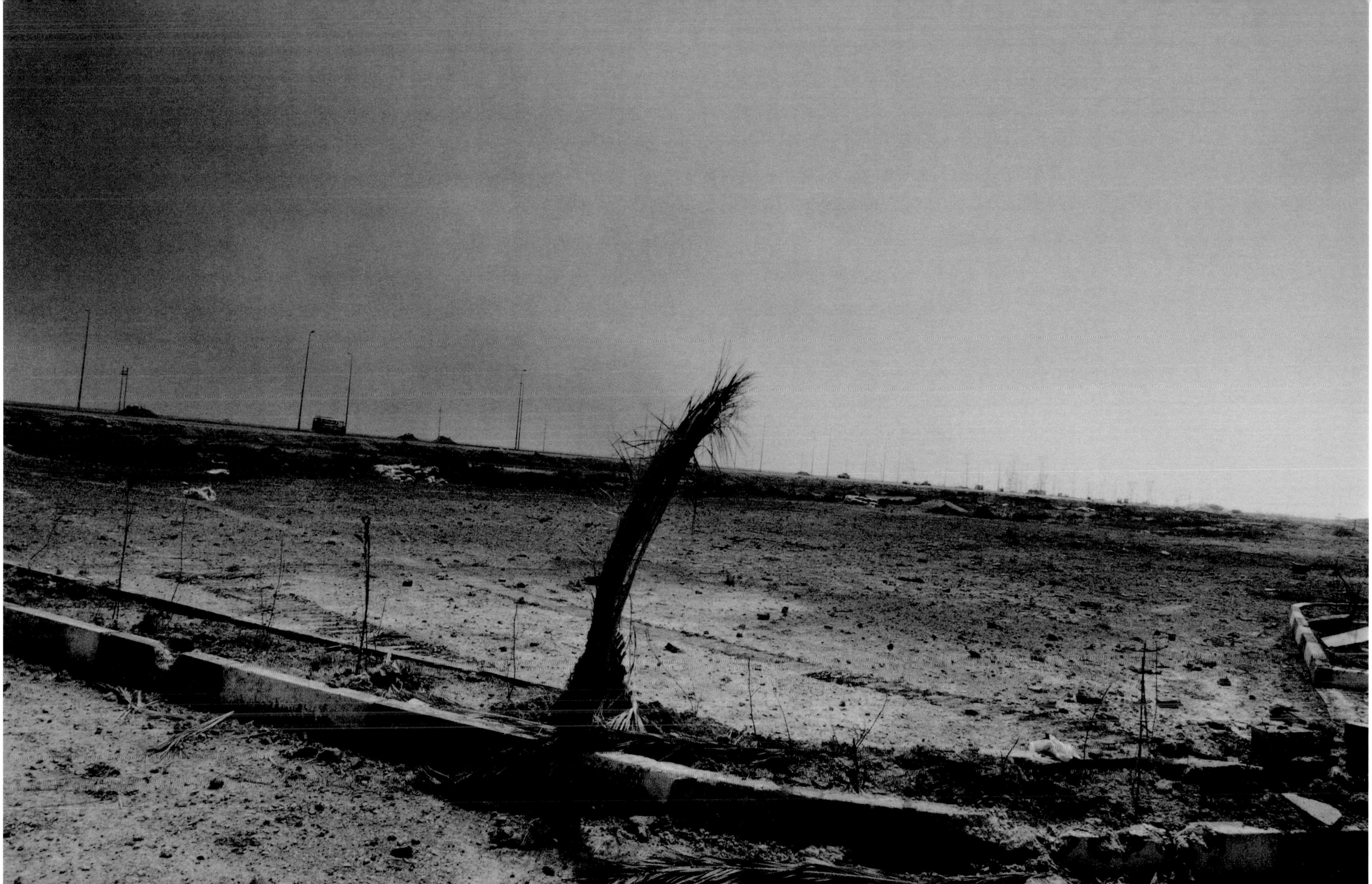

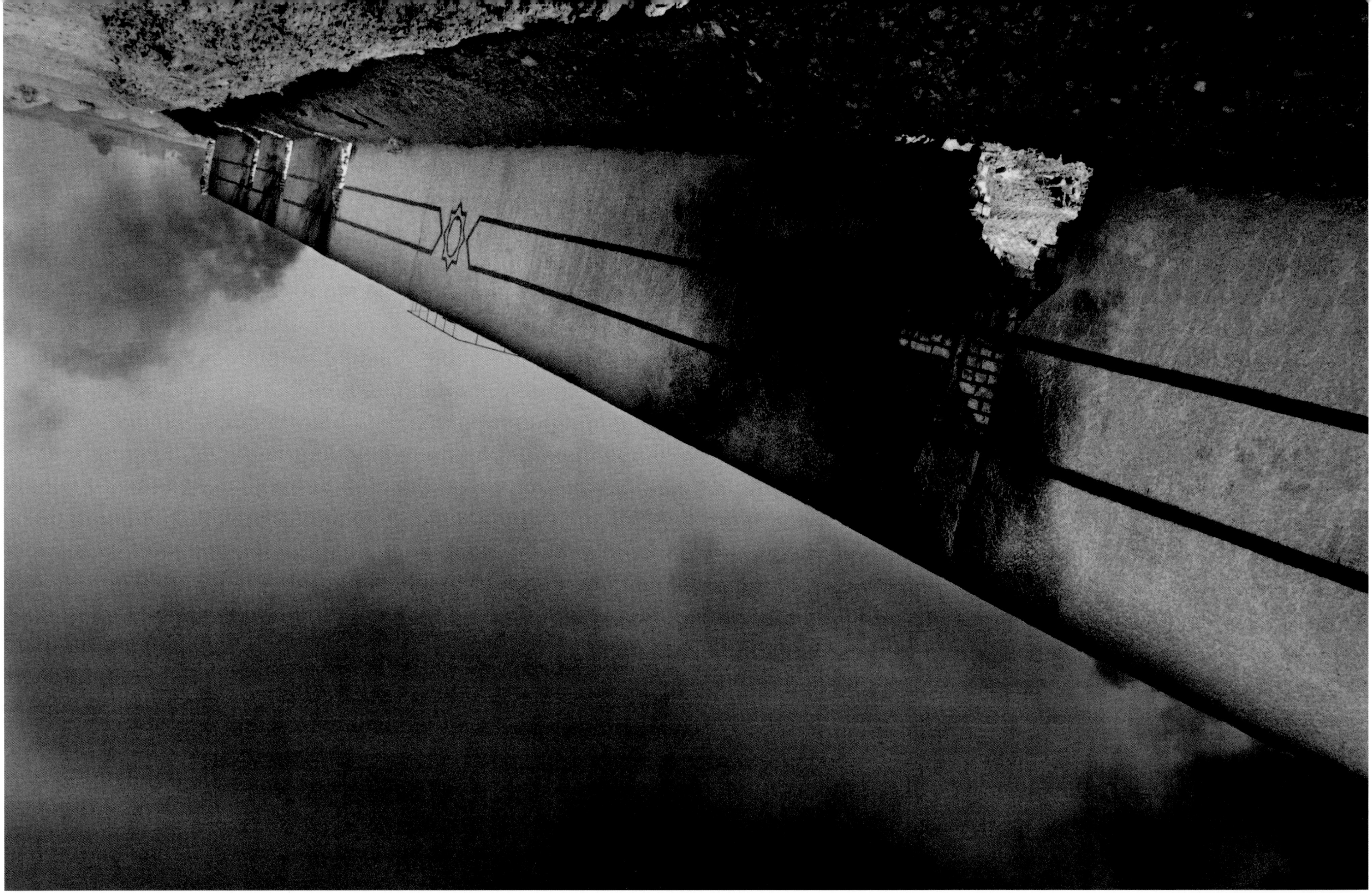

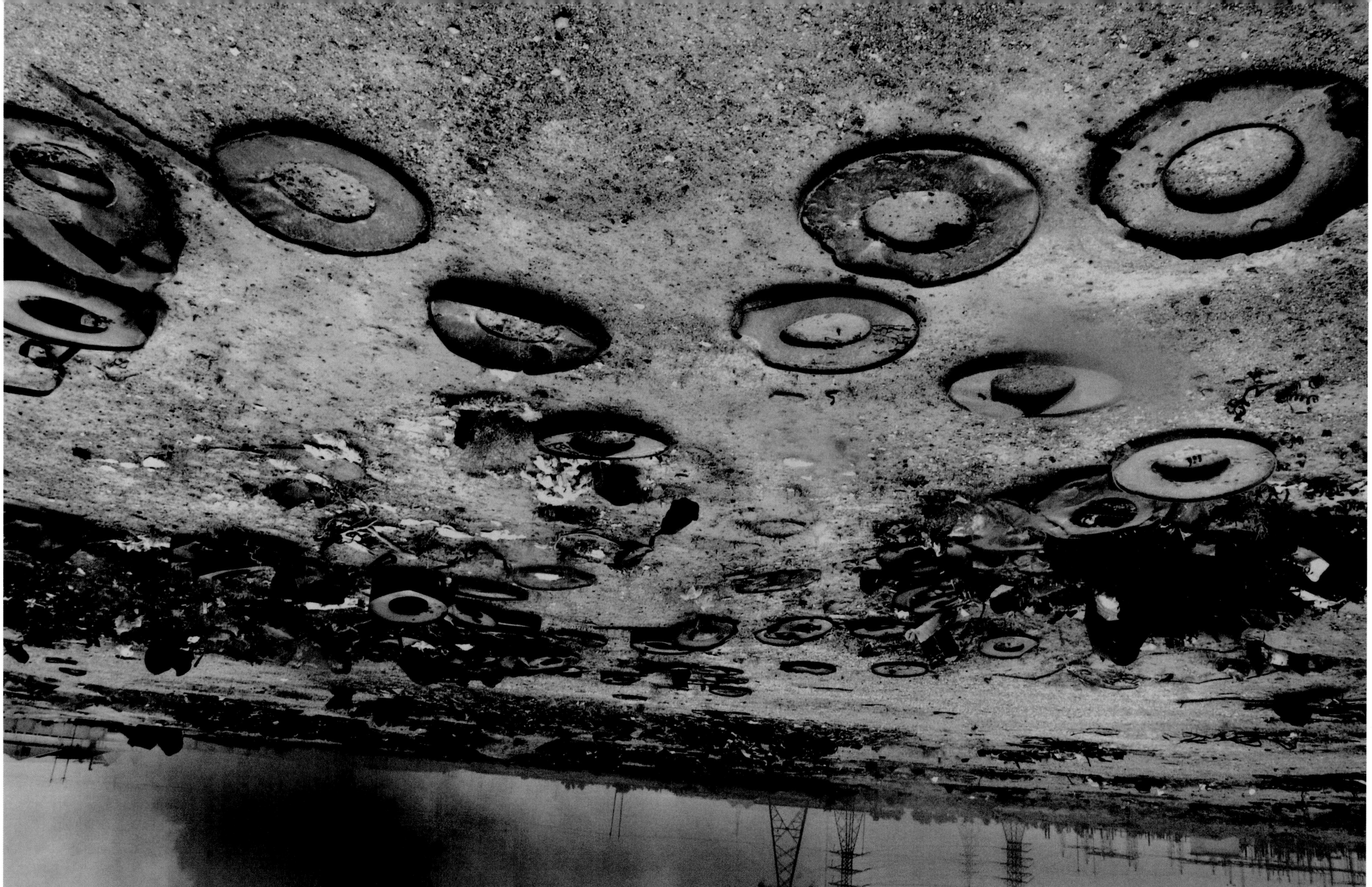

TERROR COMES IN COLORS:
yellow,
orange
and red,
and we are checked
and re-checked
again.

SUSPICION IS AN ACT OF PATRIOTIC DUTY .

A look,
a glance,
voices turned to
whispers.

People gather themselves, clutch at their presence, afraid of the slightest attention.

If you see something irregular,
report it.
If you hear anything strange,
phone it in.
If you're having impure thoughts,
dial 911.

Police stand watch at bridges and tunnels, National Guard at airports, rail and subways.
Their weapons soothing re-assurances, reminders we are safe from ... ?

It's been sown in our minds, homegrown in our heads: people want to harm us.

A Sikh was
murdered
in Mesa,
Arizona.
Folks
thought he
was
Al-Qaeda.

new york city _ homegrown fear

_ *Section*

XVI

HOMEGROWN FEAR

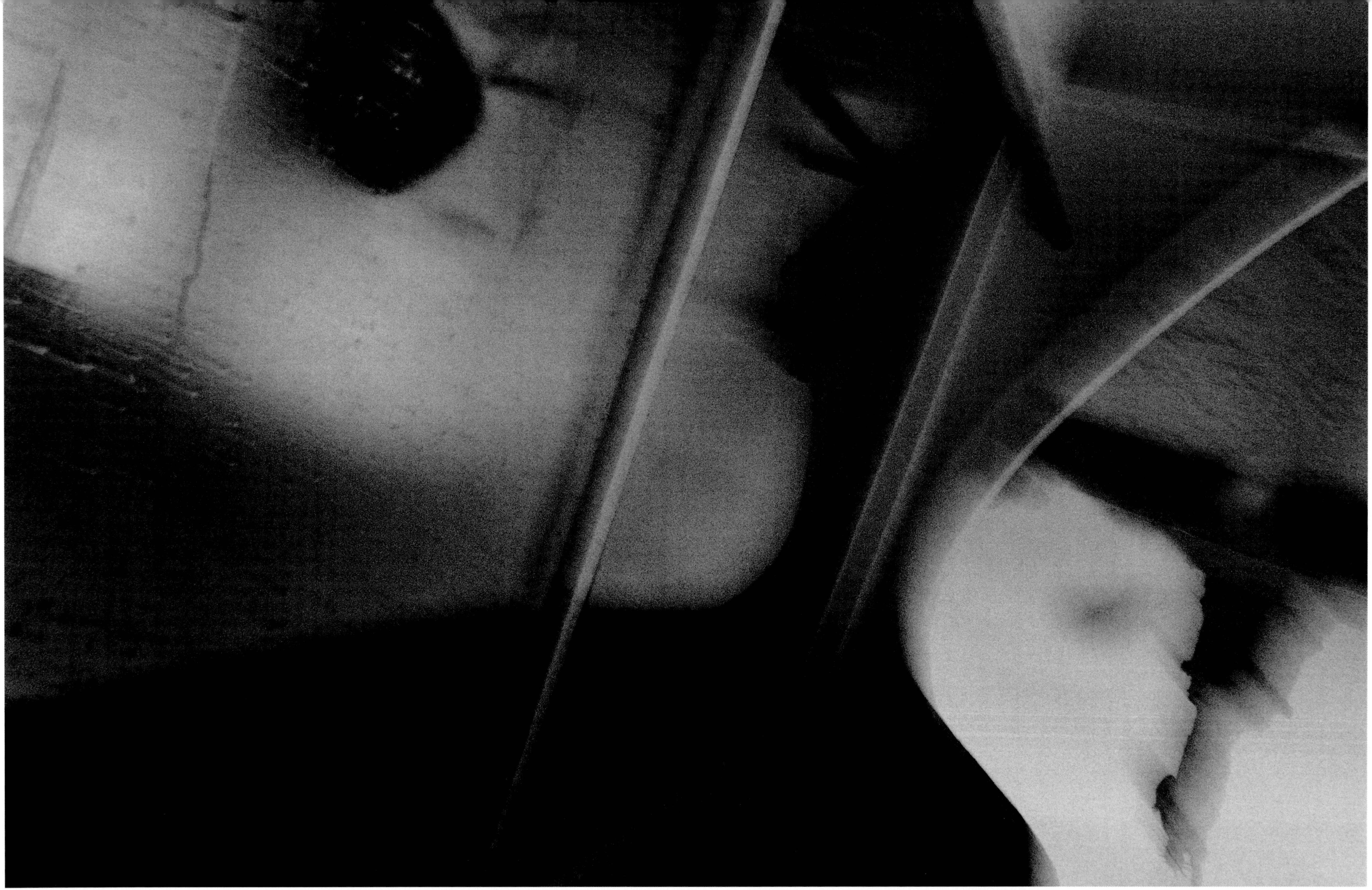

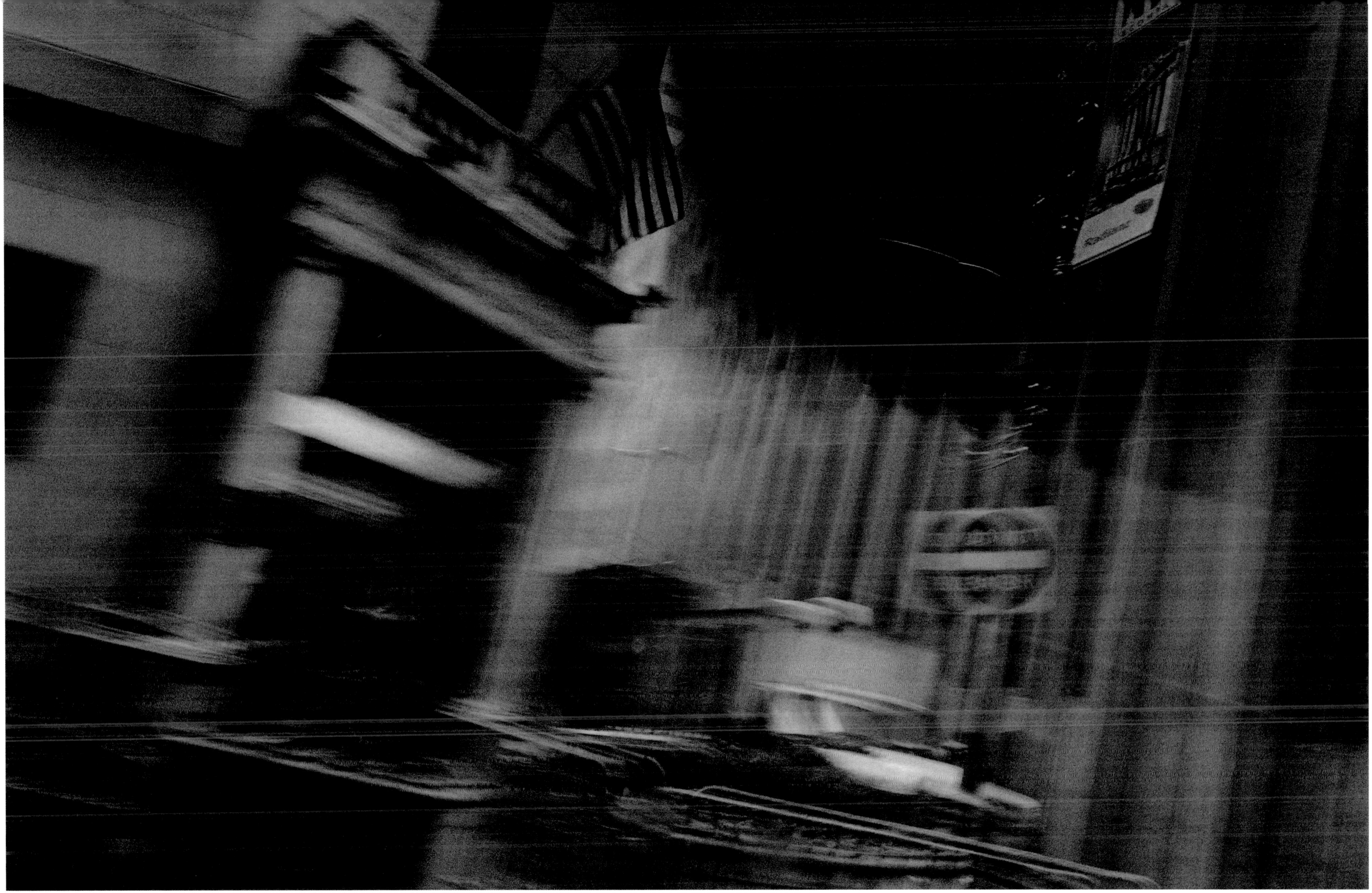

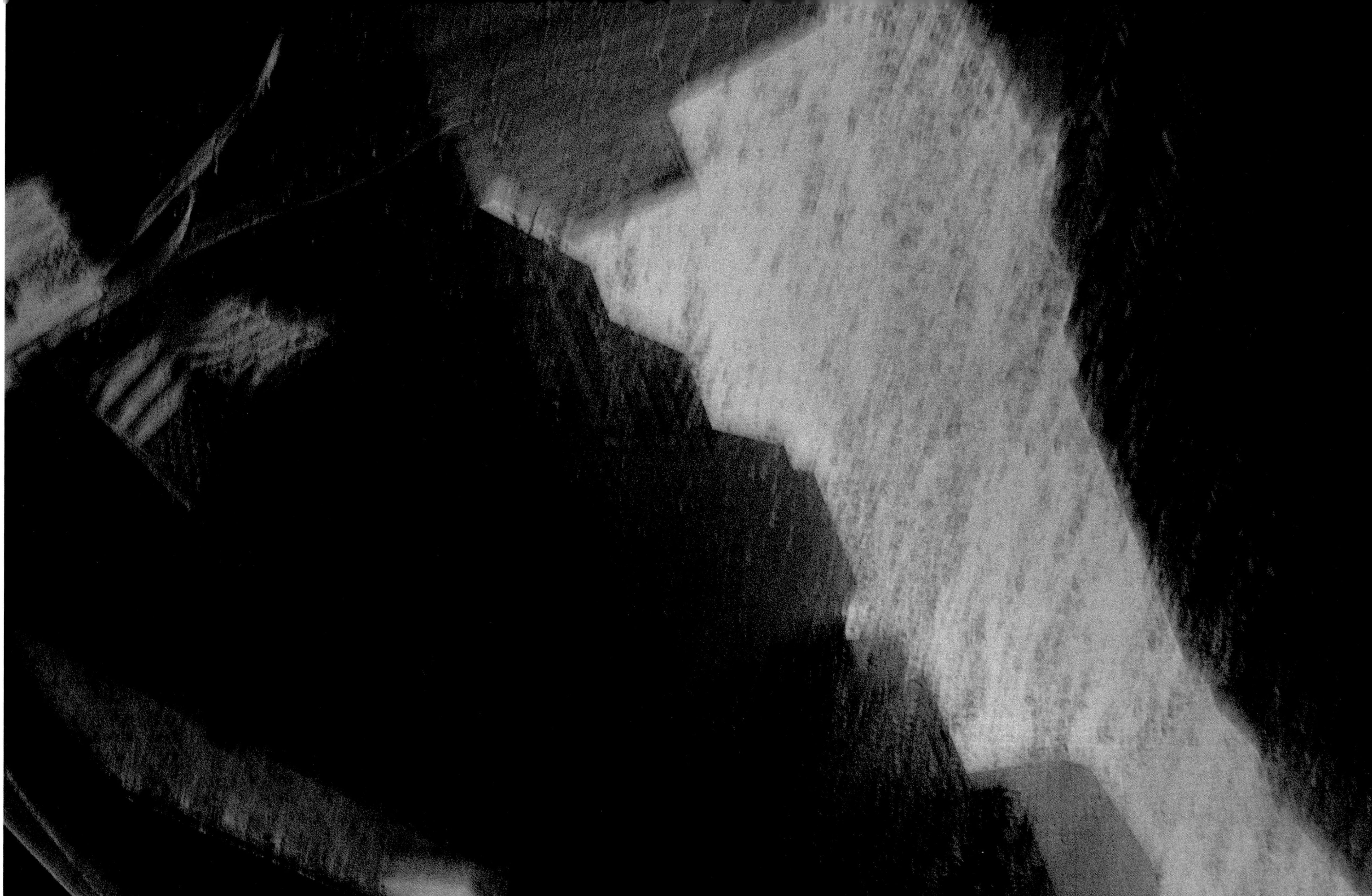

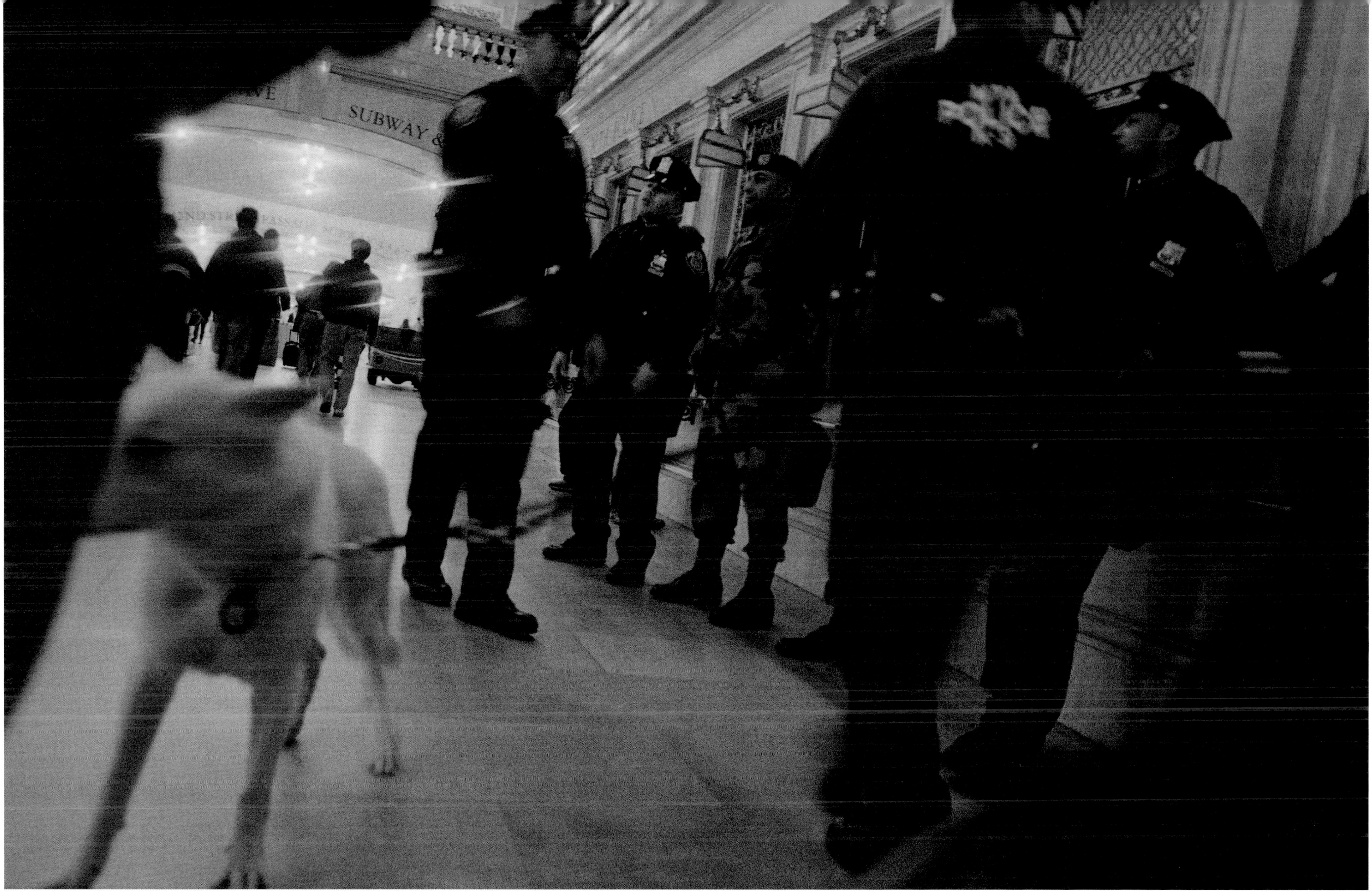
SUBWAY &

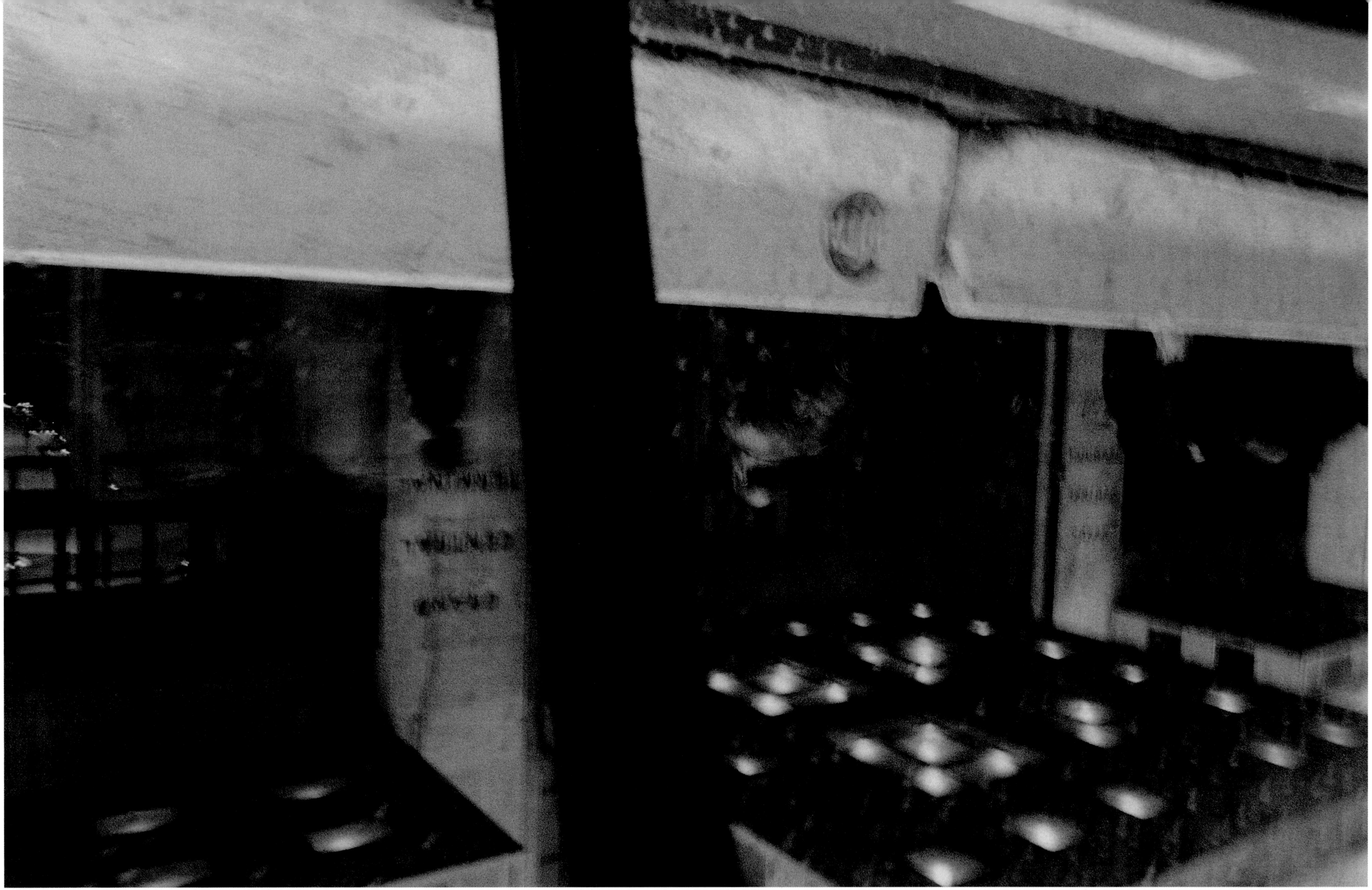

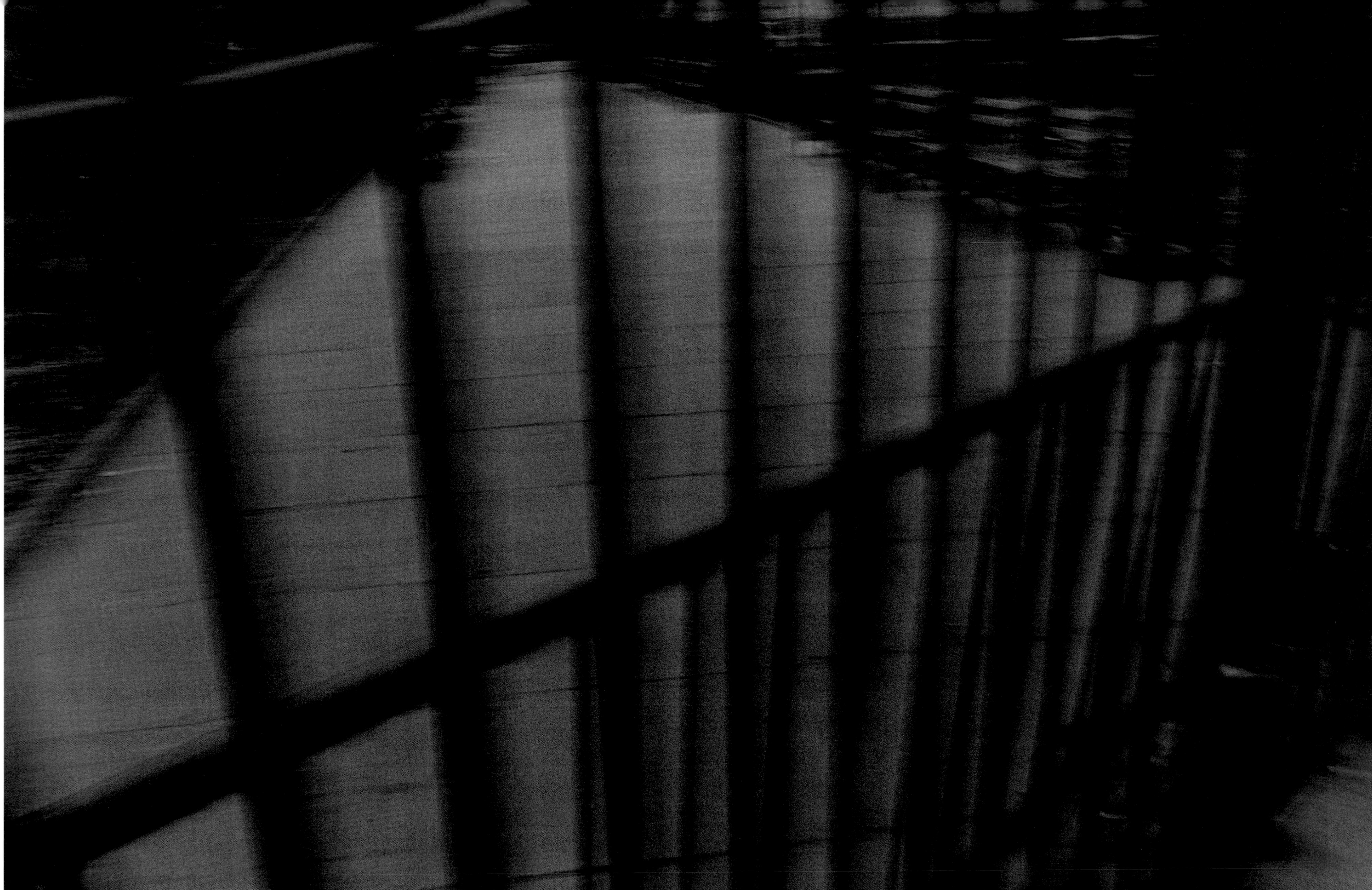

233 ENTRANCE TO MAYOR'S RESIDENCE

[VANISHING] *a book by* ***antonin kratochvil***

_ *Section* **I** : GUYANA.

Project Underground, The Gold Album, Berkeley. USA.

_ *Section* **II** : NORTHERN BOHEMIA.

Karel Kysilka, The List of Internet Sources of Czech Towns and Cities, 1997.

_ *Section* **III** : BEIRUT.

John Pike, Lebanon, Global Security.Org, 2004.

Matthew White, Secondary Wars and Atrocities of the Twentieth Century, Twentieth Century Atlas, 1999-2004.

_ *Section* **IV** : BOLIVIA.

Macelo de Urioste, The Lost Mines of Llallagua, Los Angeles Times Magazine, 1994.

_ *Section* **V** : CONGO.

Paul Clarke, The bushmeat crisis is emptying Africa's forest, The Environmental Magazine, The Environmental News Network, Inc, 2004.

_ *Section* **VI** : LOUISIANA.

Jon Bowermaster, Ghost Towns of Cancer Alley, Audubon, 1993.

Wilma Subra, Environmental Justice Communities and Industrial Facilities in Louisiana, Subra Company, New Iberia, Louisiana, 2003.

Liz Avants, A.W.A.R.E., Louisiana, 2004.

Dr. Gary Miller, Louisiana Environmental Action Network, 2004.

Ron Nixon, Toxic Gumbo, SeeingBlack.com, 2001.

Toxics, Iberville Parish/Plaquemine, Greenpeace USA, 1999.

_ *Section* **VIII** : ECUADOR.

Texaco in Ecuador, Texaco Press Release, 2003.

www.thebestofecuador.com.

Mathew Yeomans, Fool's Gold, The Village Voice, 1997.

T. Christian Miller, Suit opens against Chevron-Texaco, Los Angeles Times, 2003.

Amy Taxin, Ecuador Indians sue Texaco over polluted water, Reuters New Service, 2003.

Editors, The Fight For The Amazon, Vol. 16, Iss. 5. The Advocacy Project, 2001.

_ *Section* **IX** : PRAGUE.

Katherine Griffiths, Their jobs are on the line, The Independent Review, London, 2003.

_ *Section* **XI** : ANGOLA.

Kevin Bonsor, How Landmines Work, HowStuffWorks, Inc. Media Network, 1998-2004.

Shana Wills, Jonas Savimbi: Washington's "Freedom Fighter," Africa's "terrorist, Global Affairs Commentary," 2002.

_ *Section* **XII** : CASPIAN SEA.

Columbia Encyclopedia, Sixth Edition, 2001.

John Horvath, Cousteau Society Warns that the Caspian Sea Is Dying, Telepolis, Heise Zeitschriften Verlag GmbH & Co. KG, 1999/2000.

Energy Information Administration, Caspian Sea Region, 2001.

Ian Mather, Slashing the West's oil arteries, Scotland on Sunday, Scotsman.com, 2004.

_ *Section* **XIII** : AZERBAIJAN.

The Sturgeon of the Caspian Sea.

Elizabeth Nixon, Economics for Business 2/e, McGraw-Hill Book Company, Australia, 1999.

Traditional Garden Supply, www.scottsofstow.co.uk
Bertil Lintner, The Phuket Connection, Asia Pacific Media Services Limited, 2003.
BBC News World, Pol Pot: Life of a Tyrant, Asia-Pacific, 2000.

bibliography

Creative Director and Graphic Design _ Giorgio Baravalle . de.MO
Photo Editors _ Antonin and de.MO

Printed and bound in Italy by Grafiche Milani Segrate Milano

Published in 2005 by de.MO _ design.Method of Operation Ltd. 123 Nine Partners Lane Millbrook New York 12545

www.de-mo.org

FIRST EDITION
LIBRARY OF CONGRESS CONTROL NUMBER _ 2004108264
ISBN _ 0-9705768-3-8

© **de.MO**

Thank you for the generous support from The Victor and Erma Hasselblad Foundation.
Thanks also to Audubon Magazine, Fortune Magazine, Mother Jones Magazine, Natural History Magazine,
New York Times Magazine, Onearth Magazine, Outside Magazine.
Kathy Ryan, Kerry Tremain, Peter Howe, Michelle McNally, Jim Megargee, Rob Haggard,
Scott Thode, Kathleen Kletch, Missy May And Doug Barasch.
Special Thanks to Giorgio And Elizabeth Baravalle for bringing 16 years to life, and to
Michael Kratochvil, Michael Persson, Andrea Camuto, Mark Jacobson, Bruce Stutz, Jan Zacharias.
And Finally, my wife Gabi.

Michael wants to thank Hope Hopkins, Barbara Esstman and Tonda for another chance.

de.MO is grateful to Louise Field for her belief, enthusiasm and invaluable work for this project.
THIS BOOK WOULD NOT HAVE BEEN POSSIBLE WITHOUT THE GENEROUS SUPPORT OF DOUG TOMPKINS.

de.MO